Vehicle Painter's Notes

PETER CHILI
MIMI

BSP PROFESSIONAL BOOKS
OXFORD LONDON EDINBURGH
BOSTON PALO ALTO MELBOURNE

Copyright © 1987 by Peter Child

All rights reserved. No part of this publication may be reproduced, stored in a retrieval system, or transmitted, in any form or by any means, electronic, mechanical, photocopying, recording or otherwise without the prior permission of the copyright owner.

First published 1987

British Library
Cataloguing in Publication Data

Child, Peter
 Vehicle Painter's notes.
 1. Motor vehicles — Painting
 I. Title
 629.2'6'0288   TL275

ISBN 0-632-01873-9

BSP Professional Books
Editorial offices:
Osney Mead, Oxford OX2 0EL
 (*Orders:* Tel. 0865 240201)
8 John Street, London WC1N 2ES
23 Ainslie Place, Edinburgh EH3 6AJ
52 Beacon Street, Boston
 Massachusetts 02108, USA
667 Lytton Avenue, Palo Alto
 California 94301, USA
107 Barry Street, Carlton
 Victoria 3053, Australia

Set by Photographics,
Honiton, Devon
Printed in Great Britain by
Billings of Worcester
bound by James Burn of Eynsham

Contents

Preface, v

Acknowledgements, vi

1. Health and Safety, 1
2. Masking, 6
3. Paint Remover and Self Etch Primer, 9
4. Polyesters and Fillers, 11
5. Spraygun Method of Application, 13
6. Spraygun Maintenance, 19
7. Viscosity Cup, 27
8. Primers and Fillers, 30
9. Spraying Equipment, 32
10. Preparation, 36
11. Cellulose, 37
12. Abrasives, 39
13. What is Paint? 43
14. Polishing, 52
15. Safety, 54
16. Removal and Storage of Parts, 55
17. Power Tools, 57
18. Fibreglass, 59
19. Spraygun Setups, 62
20. Machine Polishing, 66
21. Spray Booths, 68
22. Accurate Masking, 70
23. Vehicle Wiring, 72
24. Spray Booth Coating, 73
25. Finishes, 74
26. Rust Spots, 77
27. Vehicle Manufacturers' Materials, 78
28. The BIA Research Centre at Thatcham, 81
29. Thinners, 85
30. Renovation of Second-Hand Vehicles, 87
31. Vehicle Care in the Paintshop, 89
32. Touch up, 91
33. Lining, 92
34. Industrial Fall-Out and Other Contamination, 95
35. Spraying Shapes, 100
36. Fault Finding, 102
37. Paint Thickness, 104
38. Infra-red Heat Lamps, 106
39. Spot Repairs, 110
40. Mixing Schemes, 112
41. Formulations, 116
42. Microfiche, 118
43. Basic Colours, 119
44. Colour Codes, 121
45. Basic Colour Match, 123
46. Static Electricity, 125
47. Metallic Finishes, 126
48. Colour Match in Metallics, 131
49. Metallic Spray Techniques, 136
50. Local Repairs, 137
51. Blending Clear, 139
52. Metallic Tinting, 141
53. Silicone Contamination, 142
54. Warranty, 144
55. Coatings, 145
56. Brushes, 147

Contents

57	Wet on Wet Application, 148	82	Matt Black Cellulose, 201
58	Synthetic Finish, 150	83	Paintshop Relations, 202
59	Synthetic Coach Finish, 153	84	Powder Coatings, 203
60	Hot Spray Method, 154	85	Pearl Finishes, 210
61	Lining and Signwriting, 158	86	Painting Composites (Plastic Parts), 213
62	Transfers, 161	87	Electrostatic Spraying, 216
63	Low Bake Safety Check, 162	88	Metamerism, 219
64	Oven Probes, 165	89	Fault – Bleeding, 220
65	Low Bake Ovens, 168	90	Fault – Paint Blisters, 221
66	Local Repairs (Straight Colours), 170	91	Fault – Blooming or Blushing, 226
67	Polyurethane, 172	92	Fault – Bronzing, 228
68	High Gloss Finish in Cellulose, 174	93	Fault – Cobwebbing, 229
69	Low Gloss, 177	94	Fault – Crazing, 230
70	Safety Forum, 179	95	Fault – Dry Spray, 231
71	Polyisocyanate Finishes, 180	96	Fault – Floating and Flooding in Metallic, 233
72	Material Types, 182	97	Fault – Metallic Air Pressure, 235
73	Electropriming, 184	98	Fault – Peeling, 237
74	Vehicle Washing and Care of Film, 186	99	Fault – Pinholing, 239
75	Airless Spray, 187	100	Fault – Scratch Marks, 241
76	Safety Reminders, 188	101	Fault – Sinkage, 242
77	Assessments of Time and Materials, 189	102	Fault – Solvent Popping, 244
78	Estimating, 194	103	Fault – Striping in Metallic, 246
79	Acrylic Finish, 196	104	Fault – Trapped Air Popping, 248
80	Rolls-Royce, 198		
81	Spraying Wheels, 200		

Preface

This book has been written for the painters, paint foremen, and managers who, in a high-speed and technologically advanced world, require a little guidance to develop their skills further, and overcome some of the problems associated with refinishing vehicles.

The quest for excellence proceeds and the demands on the vehicle painter will grow as the motor manufacturers develop systems and more flamboyant colours, and the owners are educated to become even more neglectful of their cars.

I have met many people in this fragmented Refinishing Industry who are dedicated to quality and excellence, and I hope that they, by example, will enthuse others, and that this book will encourage them all to renew efforts to match the challenge that the correct refinishing of vehicles brings.

Peter Child

Acknowledgements

I gratefully acknowledge all the help and kind consideration shown to me by the following, without whose help and co-operation this book could not have been written.

Aston Martin Lagonda Ltd
Ault Wiborg Paints Ltd
DeVilbiss Company Ltd
Arthur Holden & Sons Ltd
ICI Paints Division
S.Harrison Esq.
J.R.Taylor Esq.
K.Thompson Esq.
A.M.Stevens Esq.

1
Health and Safety

1.1 Fire and explosion risk
Flammable products – storage precautions.

1.2. Ventilation and extraction
Stop build-up of vapour – air flash.

1.3 Source of ignition
No matches – no smoking.

1.4 Cleanliness
Spillages – personal

1.5 Inhalation
Dust and fumes – use mask.

1.6 Skin and eye contact
Use gloves – creams – overalls.

1.7 Ingestion
No food.

1.8 Safety regulations
Must be obeyed.

1.1 Fire and explosion risk

All cellulose and acrylic finishes and their solvents and thinners come within the meaning of flammable products. Special storage precautions must be taken. A special building away from the main workshop should be constructed with easy access for the fire brigade if necessary. There should be flame-proof fittings in the store and it should be kept clean and tidy. Any spillages should be dealt with at once. Insulation against extremes of cold and heat is advisable. Technical help and assistance is available from fire inspectors and insurance companies. Sometimes refinishers make the mistake of combining paint and oil storage. This is not good practice and these items should be stored separately. The store should be brick built and have a step over so that in the event of fire the liquid contents remain trapped inside the well.

1.2–1.3 Vehicle Painter's Notes

1.2 Ventilation and extraction

Both the store and the general paint preparation shop must have adequate ventilation. Any fans used must be flame-proof and conform to the fire regulations. The position of the fan exit is important as contamination of nearby houses or adjacent buildings must be avoided. It is important to ensure that no build-up of vapours or fine dust takes place as this can ignite in the atmosphere and an air flash will result. This situation is highly dangerous and will inevitably lead to a full-scale fire. In the spraybooth air movement of at least 150 cubic feet of air per minute is the norm and this ensures that the booth is correctly evacuated of all overspray and fumes. Normally these are trapped in either a full water wash or a filter system.

Do not work in dusty and poorly ventilated areas. The practice of spraying primers and fillers in the open shop should not be encouraged. It is the duty of foremen and management as well as the shop operators to ensure good and clean working conditions.

1.3 Source of ignition

Under no circumstances must paintshop operatives be allowed to smoke in the shop. It is important that every precaution be taken to ensure that smoking is carried on in a suitable staff room, and that matches and any other lighting equipment are left in lockers. Static electricity is a very serious potential hazard and all precautions must be taken to avoid it. Paint mixing benches and preparation benches must be earthed, and vehicles must be earthed when using two-pack materials and electrostatic guns and equipment. 'No smoking' notices must be prominently displayed and under no circumstances should customers be allowed to enter the shop. Once a solvent fire has started, it is very difficult to extinguish it. Fire extinguishers should be in much evidence and should be regularly checked for condition. Factory inspectors and fire officers will advise on each situation. Vehicles should be pushed round the shop and not started up and run. Batteries should be disconnected and items such as aerosol cans should be removed before the vehicle enters the paint shop. If a low bake oven is in operation the fuel level of the vehicle should be checked. For safety it should be between a quarter full and three-quarters full. Either side of this amount can cause tank expansion due to pressurisation. Ensure spray painting operatives wear sensible safety shoes without steel clips in the soles.

1.4 Cleanliness

Within the paintshop environment cleanliness is essential, both for safety and the cleanliness of the finished work. Dirt inclusions are one of the biggest problems that paintshops encounter.

For safety at work it is essential that the preparation bays as well as the spray booths are kept completely clean. Old rags, mutton cloth, polishing mops and masking paper should be removed from the job into a suitable metal bin which should be emptied at least once a day. This refuse should be removed well away from the paintshop and preferably to a skip designed and designated for that purpose alone. All dirty and unused solvents should be disposed of into a large barrel stored away from the paintshop and removed by a company that specialises in this service. It is preferable for cleaning purposes to have the new air-driven cleaning tanks designed exclusively for paintshops. Spillages of paint either in the mixing room or on the shop floor must be cleared up immediately. Using sand or an equivalent the paint should be absorbed up and cleared away and then a rag with cleaning solvent should be used to remove all trace of the paint.

Personal cleanliness is very important and barrier creams should be used for the hands and face, and correct masks for preparation work, dry sanding, etc., and a full face mask, pressure-fed, should be used when spraying paint of any type. It is absolutely necessary when spraying two-pack materials, particularly isocyanates as these are extremely hazardous. Disposable overalls with hoods are desirable for both protection and cleanliness. Every care should be taken to promote good housekeeping within the paintshop.

Spray painters who show any signs of discomfort must seek medical advice.

1.5 Inhalation

It is vital to use proper masks when preparing vehicles for spraying and pressure-fed masks when actually spraying all paint products. The dangers associated with all types of material, especially isocyanates, are well documented. It is important to follow the paint manufacturers' recommendations when using their material. The technology of spray finishing motor vehicles today is much more advanced than earlier post-war times when only cellulose and synthetic finishes were available.

If any operator experiences any difficulty with breathing or irritation to the eyes or throat then medical advice must be sought. Many large refinishers are insisting on a medical as a condition of employment.

1.6 Skin and eye contact

It is important to use barrier creams and to take great care of the skin that is exposed to solvents and fumes. Any paint or solvent spilt on the skin must be removed at once. Thorough washing should then follow on immediately. Solvents defat the skin causing cracking and splitting, and any cut, no matter how small, can cause pain and intense irritation. This can be the base for further skin disorders. Isocyanate hardeners if spilt on the skin must be cleaned off instantly as they pose a particular hazard, in that the material can be absorbed into the skin.

Eye contact is dangerous and safety goggles or a mask should be used at all times. The eyes should be flushed out with copious amounts of clean water and medical advice sought immediately. Take no chances as the results are very serious indeed.

1.7 Ingestion

Unfortunately in the refinishing business, so much is demanded so quickly, and often there is no time to stop for lunch. A cup of coffee out of a machine and a cheese sandwich hastily eaten in the preparation shop are all there is time for. However, eating any food or drinking anything within the confines of a paintshop is hazardous. Polyester dust from body preparation and flatting dust from primers and fillers are ever present in the atmosphere, and will certainly end up on the food. These materials are extremely harmful as they cannot be broken down by the body's system, and therefore they can cause many types of disorders. A proper and well-defined staff room must be available for all staff to rest and eat food that they have brought to work. The staff room should be clean and ventilated, with proper waste bin or some method of disposing of unwanted food and drink. It is most important that management make this facility available.

1.8 Safety regulations

Every company will have safety regulations over and above the statutory legislation. These company rules must be obeyed as a consequence of not doing so is that a person may be dismissed instantly. An operator who smokes in the paintshop can expect nothing else but instant dismissal, because by his actions he endangers his fellow workers and the premises in which he

Health and Safety 1.8

works as well as the company. An insurance claim is unlikely to be passed without a full investigation and may be made null and void by this person's action.

Safety regulations are there for everyone's benefit and they should, and must, be obeyed.

2
Masking

2.1 Important factors:
Work on a clean car to eliminate dust and dirt.
Use the correct solvent-proof brown paper.
Do not use newspaper.

2.2 Masking up:
Mask out with thin tape to get an edge.
Apply the taped masking paper to the edge tape.
Be accurate and avoid bridging.
Remove masking after priming and remask.
Unmask straight after low baking.

2.1 Important factors

A. When working on any vehicle, no matter how old or how much service the vehicle has seen, you can be sure that it will have amounts of road dirt and other general grime on it. It is therefore important to wash the car thoroughly before commencing work and to blow out returns, door surrounds and drainage channels to remove as much dirt as possible at the earliest stage. When the vehicle is clean it can be masked up. It is good practice to wipe around any window rubbers that may be masked up with white spirit to remove road film and give a good surface to stick the masking tape to.

B. A properly equipped paintshop will have a masking machine with several sizes of solvent-proof brown masking paper and automatic tape dispenser supplying the roll of paper.

C. On no account use newspaper as the ink is not solvent-proof and yesterday's headlines can be transferred to your job.

2.2 Masking up

A. The correct procedure is to mask up using a small tape, up to 19 mm, and then infill with the masking paper off the dispensing machine. Mask up with the gloss side of the brown paper facing out towards you.

B. Having masked up in outline, infill carefully and ensure the finished job is completely flat with no open folds or gaps in the paper. Tape the folds down flat, because if left, they attract and

Masking 2.2

hold spray dust, and they can tear when the gun is being passed over the job. The air pressure can cause this occurrence.

C. It is very important to mask very carefully at this first outline mask up, as any tape fitting too closely to the painted surface will cause the paint being applied to 'bridge' (see Fig. 1) and this means that when the tape is removed it will pull off the fresh paint. If this does occur and the fault is realised too late after the job is finally finished then the situation can be saved if a scalpel or sharp knife is used to cut around the areas that have bridged. It is far better to avoid this by ensuring a gap between surfaces.

D. After the priming and flatting stages, and following a full wash down, the masking should be removed and the edges carefully cleaned up and the vehicle remasked for colour.

It could be that there is a 'blend out' area, in which case ensure that the whole panel is masked out.

Never apply masking tape and paper across the middle of a panel, unless it is on the edge of a body line, because no matter what, there will always be a line where the overspray finishes. Always allow the material to ghost out over the whole panel.

When the vehicle is ready for colour application remask in the normal way ensuring that all areas are blown out with the air duster to remove any dirt or sludge that may have accumulated.

When the vehicle is fully masked up and in the spray booth, after blowing off, tak ragging, and finally dusting over with the spraygun, air only, it is good practice to dust over the masking paper with a colour coat to ensure that any dust particles are stuck down to the paper surface. Give a final blow off with air before commencing to spray the colour.

When the job is complete the vehicle should be demasked as

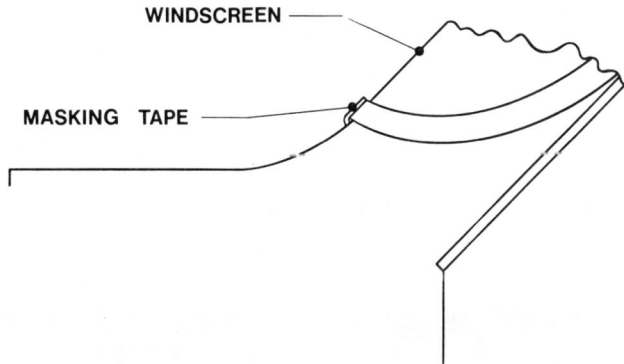

Fig. 1. Paint bridging

2.2 Vehicle Painter's Notes

soon as possible. With air dry material this can take place as soon as it is dust-free.

E. After a low bake process the masking should be removed as soon as the vehicle has cooled to an ambient shop temperature. If it is left for too long, difficulty in removing the tape may be experienced. If the job has been left overnight fully masked up, then to ease the removal of the tape a little applied heat in the oven will soften the tape and help its quick removal.

It is important not to tension tape, as it can become unstuck when heat is applied in the booth. Tape should always be firmly placed down without tension so the adhesive can give its best possible performance.

3
Paint Remover and Self Etch Primer

> **3.1 Paint remover:**
> Use water washable chemical remover with safety precautions.
> Use gloves and goggles.
> Aid the chemical action by scoring the panel.
> Wash down with warm water.
> Use only special strippers for GRP vehicles.
> Take special care with aluminium-built vehicles.
>
> **3.2 Self etch primer:**
> Used for adhesion promoting on steel, GRP and aluminium.
> Zinc chromate and acid mixed etches into panel.
> Pot life of mixed material is up to eight hours.

3.1 Paint remover

A. All paint manufacturers will offer a water-washable chemical paint remover. This material contains a wax to hold it on a vertical surface and is normally applied by brush. During application full safety procedures must be followed. It is most important to wear gloves and goggles for this operation as normally the ingredients are harmful and contain chlorine. It is very dangerous to smoke when stripping a vehicle as the fumes from the stripper, if drawn through a cigarette and heated, can turn into chlorine gas.

B. To aid the chemical stripper, light scores can be made in the paint film to help the chemical penetration into the substrate. This speeds up the job and allows the material to work more effectively. As the old film wrinkles up, remove it with a stripping knife. Let the chemical do the work and do not use unnecessary mechanical work to strip the panel. All the old film that is impregnated with stripper should be removed from the shop at once.

C. After stripping the panel, wash down with warm water to clean off the residue and finally blow dry with an air duster. To ensure a good job it is well worth wet flatting the panel with 180 wet and dry, to abrade the surface and clean off completely any remaining stripper. Ensure that all edges and channels are totally clean and that no stripper remains.

3.1–3.2 Vehicle Painter's Notes

- **D.** When removing paint from cars built with GRP such as Lotus and Reliant it is essential that a special stripper designed for GRP is used. The normal strippers will have a detrimental effect on the fibreglass and must not be used.
- **E.** When stripping an aluminium panelled car it is important to exercise great care as the stripping knife can score the panel work quite badly if proper care is not taken. This is a case where the chemical must be allowed to do its job.

3.2 Self etch primer

- **A.** Self etch primer is used on steel to obtain a good key for further coats of material. It is not essential on steel but certainly makes a better job as far as longevity is concerned. High class refinishing shops will use it every time. However, it does have to be used on fibreglass and aluminium panels. The adhesion of normal cellulose primers is not really sufficient to give the desired key. Self etch promotes adhesion, and adhesion is one of the most important words in a painter's dictionary. Without first coat adhesion the most wonderful paint job ever will flake off in service.
- **B.** By the use of zinc chromate and orthophosphoric acid mixed, an etching process into the steel, fibreglass or aluminium takes place. The acid eats into the metal structure and takes the chromate with it. Always apply one light coat only, normally approximately $\frac{1}{2}$ thou or 12.5 microns, and then coat up with normal filler coats as soon as the self etch primer has dried out, normally in about ten to fifteen minutes in a controlled spray booth. If the primer is left overnight uncoated then it will take up moisture and cause a breakdown in adhesion at a later date.
- **C.** There is a pot life of up to eight hours once the material is mixed up, and any material left at the end of the day should be discarded. After the application of filler coats over the self etch, make sure when flatting back the filler that edges are not broken through to bare metal. They can edge up later if this has occurred. To rectify this, it is possible to brush in a little self etch and cover the break and then spray filler over when the etch has dried out. It is good practice to use self etch primer on all bare metal panels and it will ensure that the paint film adheres to the metal structure of the vehicle for many years.

4
Polyesters and Fillers

4.1 Material:
Apply by brush or spray giving a high film build.
Dries out by catalyst action.
Gives a fine substrate that will not shrink or shrivel.
Material gives excellent adhesion.

4.2 Use:
Must be used on clean unpainted surface.
Can be shaped by hand rubbing or DA sander.
Make up only the amount required for the job.
Avoid heavy applications.

4.1 Material

A. The development of polyester resins over the last decade has brought to the refinishing trade materials of excellent quality and conformity. Polyester fillers can be applied by brush or spray, and most of the development work has been with spray application in mind. Generally, the material is applied from a gravity feed gun and the film build that it delivers is extremely high. This enables imperfections and weld areas to be sealed in and resurfaced with an inert material that shows no visible sign of shrinkage.

B. Polyester spray fillers are extended with talc for high build properties and catalysed with an inorganic peroxide. The material cures out by the chemical action of the catalyst which is in turn accelerated by temperature. In a spray booth operating within the temperature window of 68 to 72° F, the cure out takes between 15 and 30 minutes. This can be further accelerated by heat lamps or low bake application. Each paint manufacturer will clearly state drying times in varying conditions.

C. Polyester fillers can be flatted back to give a superb substrate for further materials that will not sink or shrivel. They are not affected by cellulose, acrylic or any two-pack solvents. They can be overpainted with any paint system. They are particularly advantageous in the low bake field as no movement of the material takes place in the substrate.

D. Adhesion levels of the material are very high and care must be taken to ensure when spraying that no other vehicle is contaminated by overspray as it is very difficult indeed to remove it from the finish of a vehicle.

4.2 Use

A. The material must be used on clean, bare scuffed metal or GRP. It should not be used over a paint edge as during flatting the paint surface is weakened and becomes more solvent sensitive. This leads to the problem of 'ringing' around the inert polyester. When a repair is made, all paint should be cleaned back to give at least $\frac{1}{2}$ inch of bare metal round the polyester.

B. The material can be shaped by hand or by the use of a DA sander. When dry sanding with discs remember that they are lubricated with stearate which if left on the job can affect adhesion. Therefore it is important to thoroughly wipe down with cleaning solvent or inter-coat wipe to clean off any residue.

C. When spraying the polyester make up only the amount required as the pot life is quite short. Do use the correct cleaner to clean the gun out otherwise it can be impossible to use the gun again without a major strip-down. Spray application should be light to begin with to set up the film and then followed by heavier application. Be careful not to over-apply as runs will result, with the usual time loss in rubbing them out.

D. Avoid too many coats of material otherwise some cracking might occur at a later date. It is important to follow the manufacturer's advice and recommendations very closely. Spray in the spray booth and always wear a full face mask when actually using the material. For fine finishing it is better to use wet and dry paper to rub down with, and for better control use a rubbing block.

5
Spraygun Method of Application

> 5.1 **Types of gun:**
> Suction feed – gravity feed – pressure feed – airless spray – electrostatic spray.
>
> 5.2 **Use:**
> Spraygun uses compressed air to atomise paint.
>
> 5.3 **Spraying:**
> Following contours to keep even distance from panel and overlap 50% to obtain obliteration.
> The objective is to end up with an even coating.
> Arcing gives lighter film build at edges.
> Application parameters.
> Service life of properly applied film.
> Good gun techniques.

5.1 Types of gun

There are five types of gun currently being used for the application of vehicle refinishing materials.
They are as follows:

> The suction feed gun
> The gravity feed gun
> The pressure feed gun
> The airless spray gun
> The electrostatic gun

A. The most famous of these is the DeVilbiss JGA suction feed gun and Fig.2 shows this unit with the principal parts. This gun is used throughout the world and has a fine reputation for excellence and reliability. The thinned material is placed in a quart pot that is held beneath the body of the gun and then air pressure draws the paint up a suction tube into the front of the gun where it is fed out into an air stream and mixed immediately in front of the gun, then atomised and taken to the workpiece by the pressure of the compressed air.

B. The gravity feed gun works by allowing the paint held in a container above the gun to fall by gravity into the front area of the gun where the mixing takes place before atomisation, and is then conveyed to the workpiece.

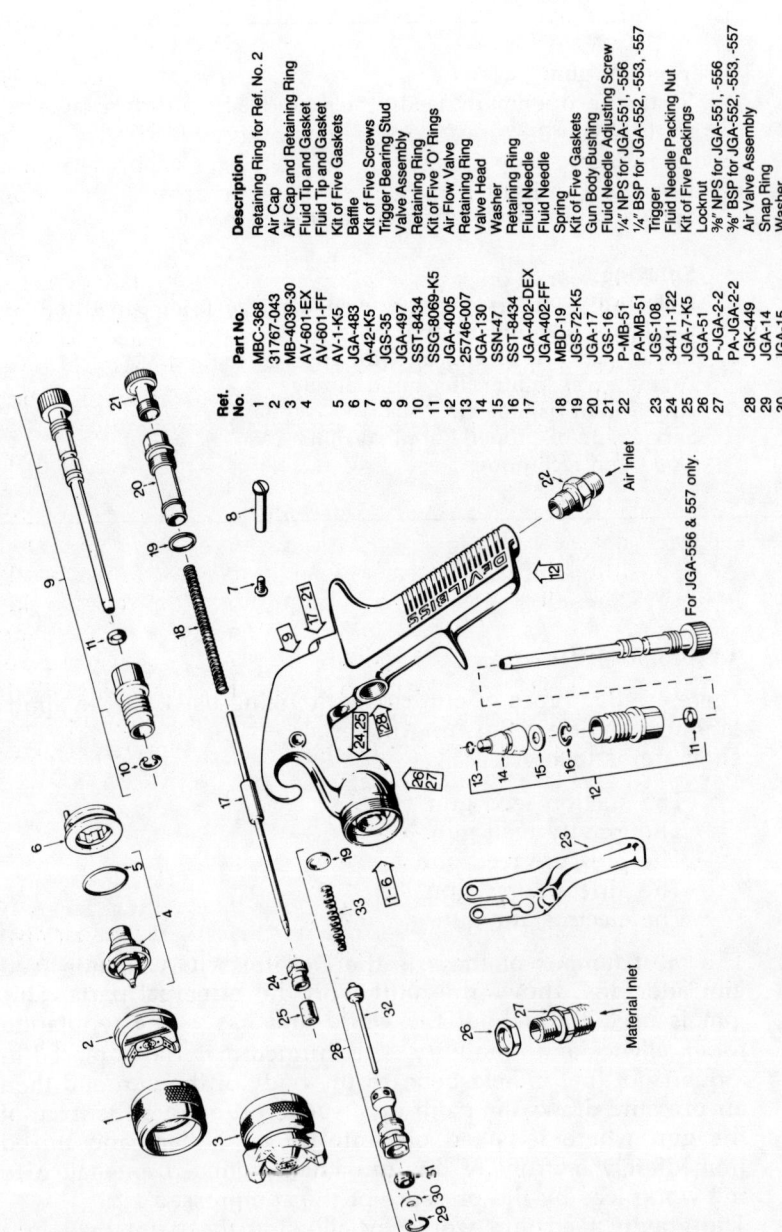

Ref. No.	Part No.	Description
1	MBC-368	Retaining Ring for Ref. No. 2
2	31767-043	Air Cap
3	MB-4039-30	Air Cap and Retaining Ring
4	AV-601-EX	Fluid Tip and Gasket
	AV-601-FF	Fluid Tip and Gasket
5	JGA-483	Kit of Five Gaskets
6	A-42-K5	Baffle
7	JGS-35	Kit of Five Screws
8	JGA-497	Trigger Bearing Stud
9	SST-8434	Valve Assembly
10	SSG-8069-K5	Retaining Ring
11	JGA-4005	Kit of Five 'O' Rings
12	25746-007	Air Flow Valve
13	JGA-130	Retaining Ring
14	SSN-47	Valve Head
15	SST-8434	Washer
16	JGA-402-DEX	Retaining Ring
17	JGA-402-FF	Fluid Needle
		Fluid Needle
18	MBD-19	Spring
19	JGS-72-K5	Kit of Five Gaskets
20	JGA-17	Gun Body Bushing
21	JGS-16	Fluid Needle Adjusting Screw
22	P-MB-51	¼" NPS for JGA-551, -556
	PA-MB-51	¼" BSP for JGA-552, -553, -557
23	JGS-108	Trigger
24	34411-122	Fluid Needle Packing Nut
25	JGA-7-K5	Kit of Five Packings
26	JGA-51	Locknut
27	P-JGA-2-2	⅜" NPS for JGA-551, -556
	PA-JGA-2-2	⅜" BSP for JGA-552, -553, -557
28	JGK-449	Air Valve Assembly
29	JGA-14	Snap Ring
30	JGA-15	Washer
31	JGS-26-K5	Kit of Five 'U' Cups
32	JGS-431	Air Valve
33	MBD-12	Spring

Fig. 2. Principal parts of a spraygun

C. With the pressure feed gun a remote pot is pressurised with the paint and that forces the paint to the underside of the gun as in the JGA set-up and paint is conveyed to the workpiece as previously described. The advantage of this arrangement is that a larger amount of paint may be in position at the job, so allowing the operator to paint large vehicles and commercial vehicles without stopping to refill. It gives flexibility to the operator allowing wet edges to be picked up easily as well as making the gun lighter and easier to use in areas such as door shuts.

D. The airless spray gun is normally used on large vehicles and buses. This works on a principle of pressurising the paint up to 3,000 PSI and then releasing it through the gun. The rapid drop to atmospheric pressure causes the paint to atomise. As a result it gives a very high build and with little overspray. If the paint is heated the pressure can be lowered to between 400 and 1,200 PSI. Further details of this procedure follow later.

E. The use of electrostatic guns has not been quite as popular as was first thought but still a number are in use with refinishers. The gun is charged electrically at the head and deposits this electrical charge into the paint. The car body is earthed and so by deposition the particles of paint are attracted to the workpiece. As a result of this a certain amount of 'wrap round' takes place and paint is attracted to the edge and back of the panel being sprayed. For metallic finishes the electrostatic gun gives a very fine and even metallic lay to the finish. This is particularly attractive to look at as well as ensuring that no 'banding' takes place. The use of electrostatic guns is described in greater detail later.

5.2 Use

All guns use compressed air supplied to them at a steady and regulated pressure for them to operate correctly. The air draws the paint through into the gun and the mix is usually carried out externally. The paint is atomised and conveyed to the workpiece.

5.3 Spraying

A. The basics of gun application are to keep the gun at an even distance, 6 to 8 inches, and to traverse the panel in an even manner at a controlled pass speed. The gun fan must then be overlapped by 50 per cent and the panel traversed back in the opposite direction. Figure 3 shows the gun distance and the correct line to follow when spraying the shape. Figure 4 shows the correct distance and the overlap.

5.3 Vehicle Painter's Notes

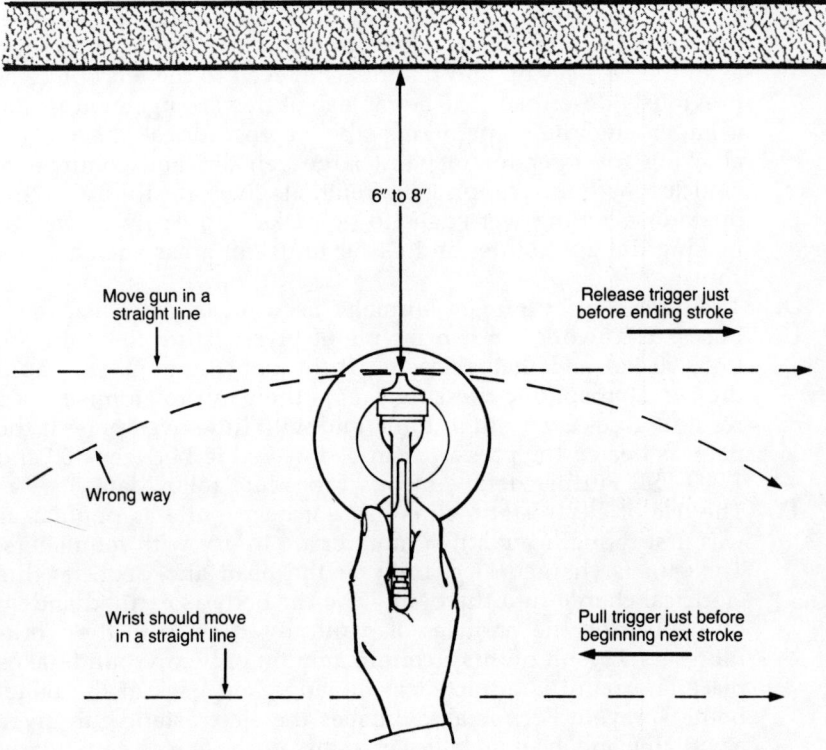

Fig. 3. Spraygun position

Figure 5 shows the effects of being too close to the panel and too far from the panel – dry spray falling on the job if the gun is positioned too far back and runs and sags if the gun is too close. Not enough has been written about gun technique and the understanding of the operation within the context of the finished job. It is important not only to practise but also to fully understand the degree of accuracy necessary to even coat a panel. The basic parameters are as follows:

- The gun must be held at an even distance, 6–8 inches.
- The gun pass must be of even traverse speed.
- The gun must be at right angles to the job.
- The gun must follow the contours of the panels exactly.
- The 'trigger off' must be in the same place every time.
- The selected 'overlap' must remain constant.

B. It is important to get all the above operations right and to keep the whole job consistent. If any deviation is made from the above a poor job in varying degrees will result. For example, incorrect overlap when spraying some metallic finishes will

Spraygun Method of Application 5.3

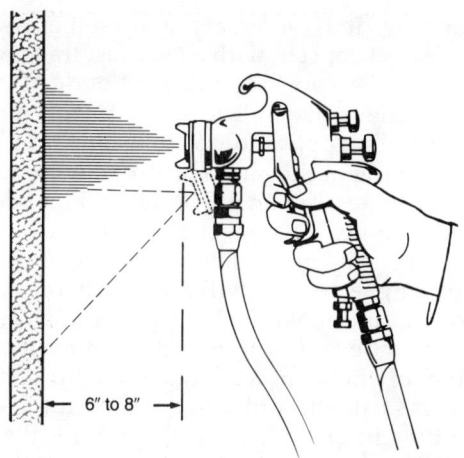

Fig. 4. Spraygun distance

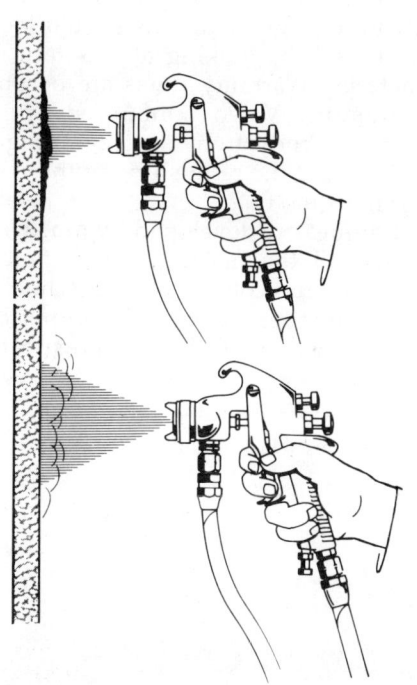

Fig. 5. Spraying fault

5.3 Vehicle Painter's Notes

cause banding. It is necessary with some metallic finishes to overlap up to 80 per cent with a very fast traverse speed. Primers and fillers can be sprayed with a 50 per cent overlap and that is quite sufficient. It is all to do with the film weight of the material and the application of that film weight.

Poor triggering at the end of a stroke will cause different film builds at the edge of a panel. Therefore door edges, for instance, may be very poorly covered, or alternatively have so much build that the paint will chip easily on the edge. In metallic finishes poor triggering technique will give a different texture and effect to the colour, and looking along the side of a vehicle then the door and panel edges will give different effects.

C. The arcing of the gun will cause a lighter film build at the extremes and will affect film build and, in the case of metallics, cause a different effect. Also dry spray results from this faulty technique.

D. If the materials are not applied within the parameters already described then the result will be an uneven film build which in the case of metallics will give a poor decorative effect as well as poor weatherability and longevity in comparison to what it could be.

E. All materials currently available will give, if properly applied, a service life of ten years or more if properly cared for in use. Refinishing work to high standards should look to this, as the OE specification will be looking at this term when first supplying the material. Warranty times are becoming longer and longer. For example, Volvo announced an eight-year warranty on paint and body from the beginning of 1986. Others will surely follow, and the refinishing of these vehicles must be of extremely high quality.

F. Good gun techniques are developed by practice and proper and detailed tuition. All the major paint manufacturers and the DeVilbiss Company run courses at their schools to demonstrate products and techniques. It is vitally important to be taught correctly by the experts, and there is no substitute for this.

6
Spraygun Maintenance

> 6.1 Regular service and maintenance of all spraying equipment.
> 6.2 Faults:
> - Fluttering spray effect.
> - Defective spray patterns.
> 6.3 Remedies for defective spray patterns.
> 6.4 Faults list:
> Excessive mist
> Runs or sags
> Orange peel
> Paint leakage
> Leakage remedy
> Air leak fault – causes.
> Oil contamination – causes.
> Overheating compressor – causes.
> 6.5 Cleaning the spray gun.
> 6.6 Gravity feed gun cleaning.
> 6.7 Care and maintenance.

6.1 Regular service and maintenance of all spraying equipment

It is important to service and maintain all the spraying equipment regularly. The spraygun is the major item to be cared for as normally in use it is most often abused. A DeVilbiss JGA gun will last indefinitely if properly maintained and cared for. Many painters make the mistake of not cleaning the gun properly and then placing it complete in a tin of solvent when it is almost unrecognisable as a precision piece of equipment. However, faults do occur but they can be quickly rectified by a good operator who knows where to look for the source of the problem.

6.2 Faults

The following list contains the causes and remedies of all the most common faults encountered in spraying.

A. Sometimes the gun will give a fluttering or jerky spray, and this is caused by (see Fig.6):
  a. Insufficient paint in the cup or pressure feed tank so that the end of the fluid tube is uncovered.
  b. When a suction feed gun is used, the cup is tilted at an

6.2 Vehicle Painter's Notes

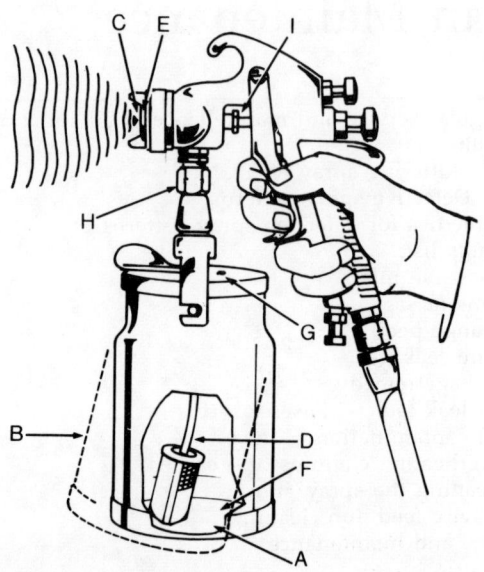

Fig. 6. Spraygun parts (faults)

excessive angle so that the fluid tube does not dip below the surface of the paint.
c. Some obstruction in the fluid passageway which must be removed.
d. Fluid tube loose or cracked or resting on the bottom of the paint container.
e. A loose fluid tip on the spray gun.
f. Too heavy a material for suction feed.
g. A clogged air vent in the cup lid.
h. Loose nut coupling the suction feed cup or fluid hose to the spray gun or pressure feed tank.
i. Loose fluid needle packing nut or dry packing.

B. The normal spray pattern produced by a correctly adjusted spray gun is shown in Fig.7, and defective spray patterns can develop from the following causes:
   a. Top- or bottom-heavy pattern (Fig.8) caused by:
     • Horn holes in air cap partially blocked.
     • Obstruction on top or bottom of fluid tip.
     • Dirt on air cap seat or fluid tip seat.
   b. Heavy right or left side pattern (Fig.9) caused by:
     • Right or left side horn hole in air cap partially clogged.
     • Dirt on right or left side of fluid tip.
   c. Heavy centre pattern (Fig.10) caused by:
     • Too low a setting of the spreader adjustment valve on the gun.

Spraygun Maintenance 6.2

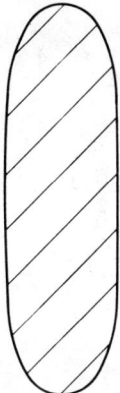

Fig. 7. Spray pattern (normal)

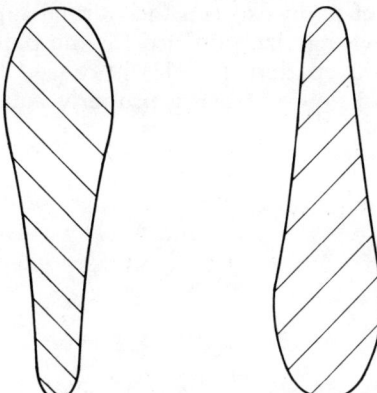

Fig. 8. Spray pattern (fault – top- and bottom-heavy)

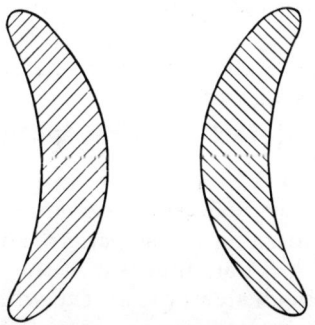

Fig. 9. Spray pattern (fault – heavy right and left side)

Fig. 10. Spray pattern (fault – heavy centre)

- Atomising air pressure is too low or the paint too thick.
- With pressure feed, the fluid pressure is too high, or the flow of paint exceeds the normal capacity of the air cap.
- The wrong size fluid tip for the paint is being sprayed.

d. Split spray pattern (Fig.11) is caused by the atomising air and fluid flow not being properly balanced.

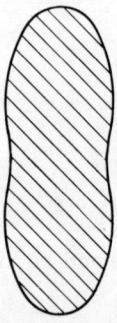

Fig. 11. Spray pattern (fault – split spray)

6.3 Remedies for defective spray patterns

For defects (a) and (b), top- or bottom-heavy pattern and heavy right or left pattern, determine whether the obstruction is in the air cap by spraying a test pattern, then rotate the air cap half a turn and spray another test. If the defect is inverted the obstruction is obviously in the air cap which should be cleaned as previously instructed.

If the defect has not changed its position the obstruction is in the fluid tip. When cleaning the fluid tip, check for a fine burr on the tip, which can be removed with 600 wet or dry paper.

To rectify defects (c) and (d), heavy centre pattern and split spray pattern, if the adjustments are unbalanced readjust the atomising air pressure, fluid pressure, and spray width control setting until the correct pattern is obtained.

6.4 Faults list

A. If there is an excessive mist or spray fog it is caused by:
- Too thin a paint
- Over-atomisation, due to using too high an atomising air pressure for the volume of paint flowing.
- Improper use of the gun, such as making incorrect strokes or holding the gun too far from the surface.

B. Runs or sags on the sprayed surface.
- Sags are the result of applying too much paint to the surface, possibly by moving the gun too slowly. Runs are caused by using too thin a paint.
- If the gun is tilted at an angle to the surface, excessive paint is applied where the pattern is closest to the surface, causing the paint to pile up and sag.

C. An 'orange peel' defect such as that sometimes obtained with cellulose and synthetic materials is caused by:
- Using unsuitable thinners.
- Either too high or too low atomising air pressure.
- Holding the gun either too far off or too close to the surface.
- The paint not being thoroughly mixed or agitated.
- Draught blowing on the surface.
- Improperly prepared surface.

D. Paint leakage from the front of the spray gun is caused by the fluid needle not seating properly due to (see Fig. 12):
   a. Worn or damaged fluid tip or needle.
   b. Lumps of dried paint or dirt lodged in the fluid tip.
   c. Fluid needle packing nut screwed up too tightly.
   d. Broken fluid needle spring.

E. Paint leakage from the fluid needle packing nut is caused by a loose packing nut or a worn or dry fluid needle packing. The packing can be lubricated with a drop or two of light oil, but fitting a new packing is strongly advised.
   Tighten the packing nut only with the fingers to prevent leakage, but not so tight as to bind the needle.

F. Compressed air leakage from the front of the gun is caused by (see Fig.13):
   a. Dirt on air valve or air valve seating.
   b. Worn or damaged air valve or air valve seating.
   c. Broken air valve spring.
   d. Sticking valve stem due to lack of lubrication.

6.4 Vehicle Painter's Notes

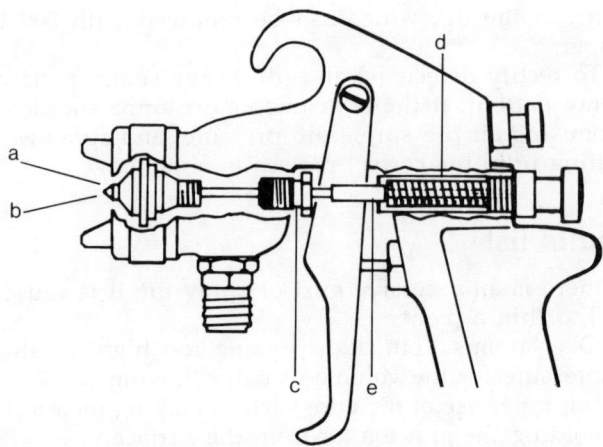

Fig. 12. Gun parts assembled

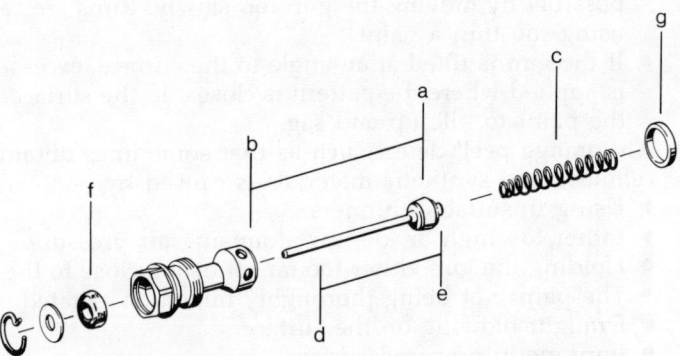

Fig. 13. Gun parts

  e. Bent valve stem.
  f. Lack of lubrication on air valve packing.
  g. Air valve gasket damaged.
G. If the air compressor pumps oil into the air line, it is for the following reasons:
  • Strainer on air intake clogged with dirt.
  • Clogged intake valve.
  • Too much oil in the crankcase.
  • Worn piston rings.
H. An over-heated compressor is caused by:
  • No oil in crankcase.
  • Oil too heavy.
  • Valves sticking or dirty and covered with carbon.
  • Insufficient air circulating around an air cooled compressor

Spraygun Maintenance 6.4–6.6

due to it being placed too close to a wall or in a confined space.
- Cylinder block and head being coated with a thick deposit of paint or dirt.
- Air inlet strainer clogged.

6.5 Cleaning the spraygun

After spraying and whilst the gun is still connected to the air line loosen the cap and, with the fluid tube still within the cup, hold a piece of rag lightly over the centre hole of the air cap. Pull the trigger and the rag pad will turn back the compressed air through the fluid tip, thus forcing any surplus paint in the gun and fluid tube back into the cup.

Empty the cup, allowing it a few moments to drain, partially refill it with a suitable cleaning fluid and re-attach it to the gun. Spray the fluid through the gun in the ordinary way, but occasionally hold the rag over the cap as before so that the cleaning fluid is surged backwards and forwards through the fluid passages, cleaning them thoroughly.

Remove the air cap of the gun and, having soaked it in cleaning fluid, scrub it with a stiff brush. If any of the holes in the cap are clogged probe them with a tooth pick or soft wood implement, but do not attempt to clean holes with a metal tool as that will irreparably damage the cap. The easiest way to dry the cap and the gun after cleaning is to hold it in a stream of compressed air.

Whilst the air cap is off the gun make sure that the outside of the fluid tip is clean of paint, and always take careful precautions to see that the tip is not damaged by knocking it while it is unprotected. When the spraygun is re-assembled wipe it clean with a rag soaked in cleaning fluid and also the outside and the inside of the fluid cup.

It is always a good plan to apply to drop of oil to the parts needing lubrication shown in Fig.14. Note that the fluid needle spring is lubricated with Vaseline or grease and this part will require only occasional attention.

6.6 Gravity feed gun cleaning

The gravity feed gun is cleaned in exactly the same way except that, of course, the cup is not detached but the lid is taken off. When blowing back with a gravity feed gun take special care to ensure that the open top of the cup is turned away from your face as a certain amount of cleaning fluid is liable to be blown out of the cup. In the lids of both gravity and suction feed cups there is a vent hole to allow air to enter the cup to replace paint

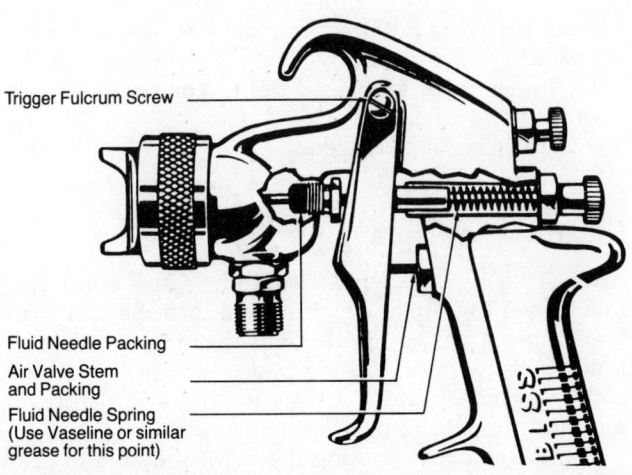

Fig. 14. Gun lubrication points

that is drawn out and it is essential that this vent hole is kept open as otherwise the paint will not flow out of the gun properly. There is also a lid gasket which must be carefully cleaned to ensure that it functions properly.

6.7 Care and maintenance

The care and maintenance of all guns and equipment are essential to the smooth running and overall efficiency of any paintshop. This aspect must not be ignored by operators, foremen or the management. Everything that can be done must be done to ensure a smooth running shop so that the true profitability of the operation can be reflected in turnover and retained profit. A clean and well maintained spraygun is part of this.

7
Viscosity Cup

> 7.1 **Viscosity cup:**
> All surface coatings are thinned and the measuring device is standard. Thinness is known as viscosity level.
>
> 7.2 **Equipment:**
> 60-second time clock, viscosity cup and stand.
>
> 7.3 **Use:**
> Paint is thinned and poured into the viscosity cup.
> The paint is released to run through the cup and the elapsed time is measured.
> All paint manufacturers give a viscosity reading.

7.1 Viscosity cup

All surface coatings are thinned to some degree or other. To measure accurately the exact viscosity of paint, primers and fillers, a specially made cup must be used. This is known as a flow cup and must be used in conjunction with a 60-second

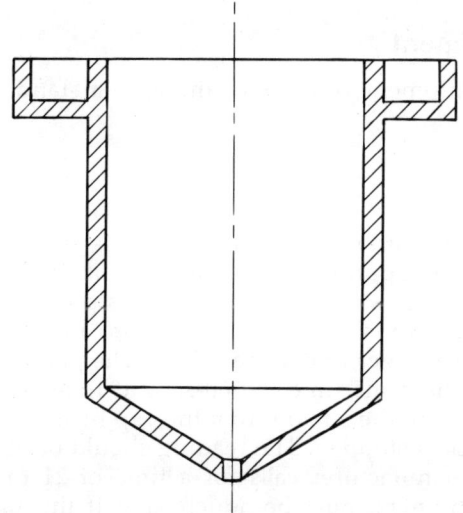

Fig. 15. Section through a viscosity cup

stop clock. Time spent thinning and checking paint viscosity can save hours of rectification work later. It is important to thin materials with the correct thinner and to the viscosity that the paint manufacturers suggest to get the best performance from the material, and it cannot be over-emphasised that thinning ratios affect colour in metallics.

The standard BSB4 is the most widely used cup and it is shown in section in Fig.15.

Ford 4 and DIN 4 are also used and the following table shows the equivalent readings:

BSB4	FORD 4	DIN 4
17	15	14
19	16	15
20	17	16
22	19	17
24	21	19
26	22	20
28	24	22
30	26	23
33	28	25
36	31	27
38	33	28
39	34	29

These readings represent elapsed seconds through the respective cups.

7.2 Equipment

The equipment consists of the cup, a stand, and a 60-second stop clock.

7.3 Use

The paint, after thorough stirring, is thinned in the case of cellulose to approximately half and half. The thinners and paint are then well stirred and a sample is poured into the viscosity cup. The cup is filled to the very top, and the usual practice is to place a finger under the orifice. The paint is released to flow and at the same time the 60-second time clock is started. The point where the stream from the cup breaks is the moment that the clock is stopped. The reading should be about 24 secs BSB4. If the manufacturer calls for a time of 21 to 23 seconds then more thinners must be added, and if the call is for 26 to 28 seconds then more unthinned paint must be added. General

Viscosity Cup 7.3

usage soon makes it apparent what is required to get very close to the required reading.

Always ensure that the temperature of the paint is correct, as cold as well as hot paint will give different readings. The paint manufacturer normally calls for about 20° C as a standard.

Always use the viscosity cup as it will ensure that one of the unknowns is taken out of the paint equation. The more professional and accurate the operator is, the less chance of errors and paint faults creeping in.

All paint manufacturers give a viscosity cup reading for every product currently on offer.

8
Primers and Fillers

> **8.1 Object:**
> To provide a key or good adhesion to subsequent coats of material.
> To contribute to the rust inhibiting properties of the paint system.
> To produce an adequate build up and smooth surface for the final colour coats.
>
> **8.2 Types:**
> Self etch primer – chip-resistant primer – fillers – primer surfacer – stoppers – non-sand undercoats.

8.1 Object

The object of all primers is to obtain a key, which will then promote good adhesion. This can apply to bare metal, aluminium, glass reinforced polyester, or an original painted surface. Where they are used over bare metal they will improve the resistance to corrosion and will have rust-inhibiting properties. Their main function is to give a secure base to the whole system. Primers are not formulated for flatting or scratch filling properties, and any rubbing down should be limited to a light de-nib.

8.2 Types

A. *Self etch primer.* A two-part material consisting of a zinc chromate and orthophosphoric acid thinner. Mixed 50–50 and requiring a single coat application of about ½ a thou or 12 microns. Do not flat or de-nib and coat up with filler as soon as it is dry. Normally about fifteen minutes in a booth with 150 cubic feet of air per minute circulation and at a temperature of 68 to 72° F. This primer must be used on aluminium and GRP. When used on steel a better and longer lasting job ensues.

B. *Chip-resistant primer.* Specially formulated for application to areas subject to stone chipping. Sills and aprons as well as air dams, etc. It gives improved resistance against corrosion and helps to reduce drumming noise.

C. *Fillers.* These are designed to fill up imperfections of body work and it is possible to spray thicker coats of filler. Make sure

Primers and Fillers 8.2

that the correct thinner is used and follow the manufacturers' specifications for application and recoat times and temperatures. Over-application can and usually does cause problems later of sinkage or solvent trapping. The polyester two-pack fillers that cure out by catalyst action of the peroxide hardener may have heavier application levels as they become inert by the action of the catalyst.

After drying out, an application of a black guide coat prior to wet or dry flatting will give the operator the clear guide to the flatting procedure, and will ensure that no area is missed.

D. *Primer surfacer.* Primer surfacers and primer fillers have two functions, to provide a degree of filling and to promote adhesion. For better filling and easier flatting primer surfacers and primer fillers contain much more pigment than primers. This high pigment level leads to poor water resistance should the material be applied too thinly. It is important to spray at least two good coats giving a minimum build of 0.002 inch or 50 microns.

E. *Stoppers.* These are usually cellulose-based materials that are applied over flatted filler coats where small indentations such as file marks are still present. They should always be applied sparingly as they tend to sink if over-applied, especially if the vehicle is placed in a low bake oven. Several coats of stopper are best applied with inter-coat times of at least an hour or more.

F. *Non-sand undercoats.* These undercoats are designed to flow out and minimise flatting. They are normally used as sealer coats immediately prior to the application of colour coats. This permits the substrate to be flatted with a coarser grade paper, whilst still offering good scratch filling properties and hence good gloss levels from the colour coats.

Non-sand undercoats may be in the form of non-sand primers, giving both good adhesion to bare metal and good corrosion resistance. Alternatively primer fillers can be formulated for use either as normal or non-sand undercoats.

9
Spraying Equipment

9.1 The equipment:
Good equipment, properly installed and working correctly, is the necessary foundation before vehicle spraying can commence. When setting up a paintshop expert advice is needed and this should be sought at the outset.

9.2 The compressor:
This is the unit, normally electrically driven, that compresses air and delivers it to the transformer.

9.3 The air transformer or regulator:
This unit receives air from the compressor and removes any oil or moisture and then regulates the pressure before delivering it forward.

9.4 The air hose:
The correct air hose delivers the air to the gun. Careful account must be taken as to the length and bore size as these factors will affect the pressure delivered to the gun head.

9.1 The equipment

It is essential when setting up a paintshop that proper advice be sought with regard to the equipment. It is a specialist field to advise exactly on the compressor size to be used in conjunction with the equipment and tools that will be used in the shop. Too many refinishing shops run with poor, well used and often unreliable equipment. To understand fully the spray painting operation it is as well to start with the basic items of the necessary equipment.

A. *The air compressor.* Compressed air is needed to commence and this unit is usually an electrically driven unit with a storage tank.
B. *The air transformer.* It is necessary to condition the compressed air for the spraygun, and this conditioning means filtering out dirt, oil or moisture as well as reducing the pressure of the air at the spraygun. This is normally done with a filter and a regulator combination.
C. *The air hose.* Clean dry compressed air regulated to the required pressure is delivered to the spraygun by means of a hose.

D. *The spraygun.* The spraygun, although coming at the end of the system, is really the heart of that system. It is the tool with which the paint is actually applied to the surface.

9.2 The compressor

The compressor, as the name implies, takes air in at atmospheric pressure, compresses it and stores it in an air receiver at a very much higher pressure. The compressor must be able to maintain a steady pressure for long periods of time without developing too much heat and delivering hot compressed air.

The very action of compressing air generates heat in itself, but in a properly designed compressor this generated heat is dissipated and the compressed air is delivered at the normal atmospheric temperature. An air receiver is used to store the air for a cooling off period, and its capacity should be in proper proportion to the size of the compressor. In cooling the air the receiver permits the moisture and all the vapour entrained in the compressed air to condense, in which state they can be easily separated from the air. The air receiver also completely blankets out pulsations from the compressor so that the compressed air is drawn from the receiver at a steady and even pressure. The air pressure required to give an effective finish with a spraygun varies with the painting material used, and to a certain extent the spraygun or spray equipment used. Air pressure is very widely understood and appreciated, in fact the pressure gauge which is fitted to an air compressor makes certain of that, but air volume, which is equally important, is very often completely overlooked until the lack of it manifests itself by falling pressure whilst spraying.

Another very important point is that compressed air for the spraygun must be absolutely clean and free from moisture, dirt and oil, any of which can ruin a sprayed surface. A standard air compressing plant for spray painting work is normally fitted with an air intake filter to prevent dust entering the compressor, and also a compressed air filter is usually incorporated in an air transformer.

9.3 The air transformer or regulator

An air transformer consists of two units, a condenser or filter for removing oil and moisture from the air, and a pressure regulator. The condenser allows the air to expand into a chamber thus cooling the air, and then removes the oil and moisture by means of an easily removable filter. A cock is fitted to draw off the accumulated impurities at suitable intervals, normally once a day.

The regulator is actually a reducing valve which regulates the air pressure from the compressor to that required for spraying. A transformer should be fitted with a gauge giving an accurate reading of the pressure of air which has passed through the regulating valve. In circumstances where exceptionally humid air is compressed and moisture tends to settle out in the air hose between the spraygun and the regulator, a small lightweight Dryit filter can be fitted directly to the handle of the gun as a final safeguard against water reaching the spray work. Also a small pressure gauge can be fitted at the spraygun handle to give the pressure of air directly at the gun.

9.4 The air hose

Compressed air is brought to the spraygun from the air compressor through a suitable length of pressure-resisting hose, and it should be remembered that the internal diameter of the hose and the overall length of the air hose used have a marked effect on the efficient performance of the spraygun and the quality of the coating that it produces.

Proper hose sizes for compressed air are just as important as the correct size wire for electricity or the proper piping for water. Often operators believe they are using an extremely high air pressure, but investigation reveals that, due to a small hose size or an extra long length, pressure is inadequate for proper atomisation. The interior wall of air hose is smooth, but even so it creates a certain resistance to the flow of air, particularly when a long length is used.

Air hose is available in a range of bore sizes and Fig.16 shows the spraying equipment and pressure drop to be expected from various lengths of $\frac{1}{4}$ inch and $\frac{5}{16}$ inch air hose when used with a spraygun consuming 12 cubic ft of air per minute at 60 lb per square inch pressure.

It is important to maintain all spraying equipment and sprayguns. Compressors should be drained daily at the end of the work period, and the drainage cocks left open so that all moisture can draw off overnight. Drive belts on the motor should be checked as per the manufacturers' instructions and oil in the crankcase should be checked at regular intervals.

Ensure that fresh clean air surrounds the compressor intake and do not place the compressor in dusty or wet surroundings. The air regulator should also be drained daily and the cock left open to exhaust all the air in the system. If a filter is fitted then this must be changed at regular operating intervals bearing in mind the daily use of that particular unit. Air hoses should be wound up and stored off the floor, and should be wiped clean after use. A great deal of dust is attracted to the air line which in use transfers itself to the job, particularly when leaning across a roof or bonnet panel. Discipline and good housekeeping make

Spraying Equipment 9.4

MINIMUM PIPE SIZE RECOMMENDATIONS*

Compressing Outfit		Main Air Line	
Size	Capacity	Length	Size
1½ & 2 H.P.	6 to 9 C.F.M.	Over 50 ft.	¾"
3 & 5 H.P.	12 to 20 C.F.M.	Up to 200 ft. Over 200 ft.	¾" 1"
5 to 10 H.P.	20 to 40 C.F.M.	Up to 100 ft. Over 100 to 200 ft. Over 200 ft.	¾" 1" 1¼"
10 to 15 H.P.	40 to 60 C.F.M.	Up to 100 ft. Over 100 to 200 ft. Over 200 ft.	1" 1¼" 1½"

*Piping should be as direct as possible. If a large number of fittings are used, large size pipe should be installed to help overcome excessive pressure drop.

TABLE OF AIR PRESSURE DROP

Size of Air Hose (I.D.)	Air Pressure Drop @ Spray Gun			
	5 Ft. Length	15 Ft. Length	25 Ft. Length	50 Ft. Length
¼ Inch	P.S.I.G.	P.S.I.G.	P.S.I.G.	P.S.I.G.
@ 40 PSIG	0.4	7.5	10.5	16.0
@ 60 PSIG	4.5	9.5	13.0	20.5
@ 80 PSIG	5.5	11.5	16.0	25.0
⁵⁄₁₆ Inch				
@ 40 PSIG	0.5	1.5	2.5	4.0
@ 60 PSIG	1.0	3.0	4.0	6.0
@ 80 PSIG	1.5	3.0	4.0	8.0
⅜ Inch				
@ 40 PSIG	1.0	1.0	2.0	3.5
@ 60 PSIG	1.5	2.0	3.0	5.0
@ 80 PSIG	2.5	3.0	4.0	6.0

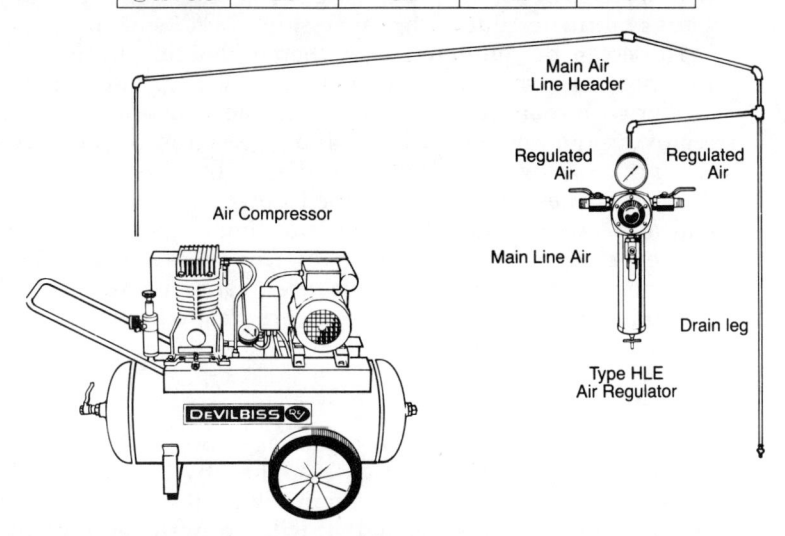

Fig. 16. Air layout and pressure table

a clean, efficient and safe sprayshop. You can never be too clean and too careful. Well maintained equipment will give a very long working life without any problem and will enhance the standard of good work.

10
Preparation

> **10.1 Metal treatments:**
> All metal panels will suffer from some form of contamination, e.g. rust or grease.
>
> **10.2 Cleaners:**
> All the major paint manufacturers offer a form of phosphoric acid based metal treatment. After correct use the metal surface is ready to accept primer.

10.1 Metal treatments

When working on new or stripped panels certain procedures should be followed to enhance the total film build and produce high quality refinishing. All mild steel and aluminium panels will suffer one of the following: rust contamination, oil or grease, alkali residues, phosphates, etc., and acid runs.

These can be countered both chemically and physically. For instance, rust scale can be removed by a grinder or a DA sander with coarse discs. This is often the most effective way of removing this contamination. Oil or grease can be removed by wiping down with a cloth soaked in solvent or white spirit. Alkali residues can also be wiped down with a solvent cloth and acid can be diluted out by washing in hot soapy water. However, after removing the worst of these contaminates a treatment with a prescribed metal pre-treatment will ensure that the grain boundaries of the metal are cleaned out chemically.

10.2 Cleaners

All the major paint manufacturers offer a phosphoric acid based metal treatment. This material is applied by brush or cloth to the surface and left to work chemically on the metal. After ten or fifteen minutes it is washed off with hot water and the panel is then dried off using compressed air. The surface is then clean and 'open' to accept a coat of primer.

This process is very important as it contributes to the overall length of life that a painted panel will offer.

11
Cellulose

11.1 Description:
A lacquer dries by solvent evaporation. No chemical change involved. High solvent content. Rapid initial dry. Blend of nitrocellulose and synthetic. Gives high build. Gloss from the gun needs 16 hours for through dry.

11.2 Uses:
At present widely used in repair industry. Can be reworked easily, polished. Very versatile.

11.1 Description

Cellulose is a lacquer that dries out by solvent evaporation, and no chemical change takes place. The material has a high solvent content with a rapid initial dry out. It is soluble in its own solvent which means that when further coats of material are applied they soften up the substrate. This fact, along with the solvent evaporation dry out, leads to heavy sinkage.

Cellulose was first used in 1928 in the UK when Sir Herbert Austin painted an Austin 7 with Cranco, a black cellulose supplied by Sir Frederick Crane Ltd. It caused a revolution, as up until that time vehicles had been finished in synthetic which was slow to dry. The material today is very robust and gives excellent gloss, and good build and hold out. The material when allowed to air dry needs 16 hours to through dry before polishing. If polishing is carried out too quickly then the appearance will go flat.

The thinners and the material are carefully matched to give the best possible results when sprayed correctly. Here is the formulation for a cellulose colour and the correct balanced thinner formulation:

$\frac{1}{2}$ sec nitrocellulose (spirit damped)	9.7
15–20 sec nitrocel (spirit damped)	0.5
Medium oil length non-drying alkyd at 50 per cent	12.6
Rutile titanium oxide	4.8
Solvent	72.4
	100.0

11.1–11.2 Vehicle Painter's Notes

Solvent blend	
Methyl isobutyl ketone (MIBK)	10
Isopropanol	15
Cellosolve	15
Methyl ethyl ketone (MEK)	30
Xylene	20
Toluene	10
	100

11.2 Uses

Cellulose finish is the most widely used material at present, but it is fast becoming replaced by the new high build two-pack materials which offer many advantages to the low bake oven user or main agent who has a very high throughput.

However, many small refinishers will carry on using this material as it is versatile in the repair of motor vehicles, and it can be interfered with on quite a scale, so that dirt inclusions, runs and sags can be flatted out and then polished, or the material can be repainted as often as the refinisher wishes without serious defects occurring. Excessive film builds will cause problems eventually, and should be avoided.

Colour matching and tinting is relatively easy, and the product has been with the industry ever since 1928. It will eventually be replaced, but exactly when is not totally sure.

12
Abrasives

12.1 Use:
To flat back surfaces, either original or filled, to obtain a key for inter-coat adhesion. Very important for the paint film durability.

12.2 Feather edging:
A flatting operation to expose bare metal where stone chips or body repairs have taken place.

12.3 Wet flatting:
Cuts freely and avoids dust. It gives good control and is usually carried out by hand.
Methods of flatting and blocking out.

12.4 Dry flatting:
Fast method of preparation and often used as first work.

12.1 Use

Abrasive papers are used primarily to flat back a surface to offer a key for subsequent coats or to level a surface prior to recoating with materials.

The papers currently in use in the UK conform to an international standard. The FEPA 'P' grades are as follows:

P 100	P 240	P 500
P 120	P 280	P 600
P 150	P 320	P 800
P 180	P 360	P 1000
P 220	P 400	P 1200

P 100 is the most coarse grit, right up to the very fine P 1200 that is used for flatting colour before extensive polishing. These grits and papers are designed to be used wet or dry, and generally vehicles are wet flatted. There are some advantages as wet flatting controls dust, cuts freely without clogging, gives good control to the operator and allows a soluble soap to be used which will help reduce grease and dirt contamination.

It is very important to select the correct grade of paper for the job in hand. Too coarse a grit will cause scratching which will show in the final colour, and conversely too fine will only 'polish' a surface that needs to be cut back. Too often the operator's favourite grade is used across the board and this is not good practice. Be selective and follow paint manufacturers' recommendations for grades of paper to be used.

More and more refinishers are now flatting 'dry' and using DA sanders with self-adhesive discs to carry out the flatting and sanding operation. There are advantages here, as speed improves and a great deal of time is saved, particularly on big jobs. However, the sander does not have the 'feel' and the operator has lost the surface contact that he has when hand flatting. It is possible to cut too deeply into a surface, and if a small particle gets trapped under the disc then this will tear up the surface being flatted quite badly. The use of stearate powder as a lubricant can cause adhesion problems if it is not cleaned away thoroughly before painting. Dry working is gaining more and more popularity and in general refinishing it is likely to become the norm.

12.2 Feather edging

DA sanding is very useful for first cut and particularly where an edge needs to be 'feathered'. To feather edge correctly is vital to a good surface finish. Care must be exercised here, and the trap is that operators believe that the superior filling qualities of fillers today make feather edging almost a thing of the past. It is not so, as no matter how good the filler, it is asking too much to cover what is in effect a paint fracture. The edges must be properly feathered back, as in Fig.17, to give a good substrate to the following paint system.

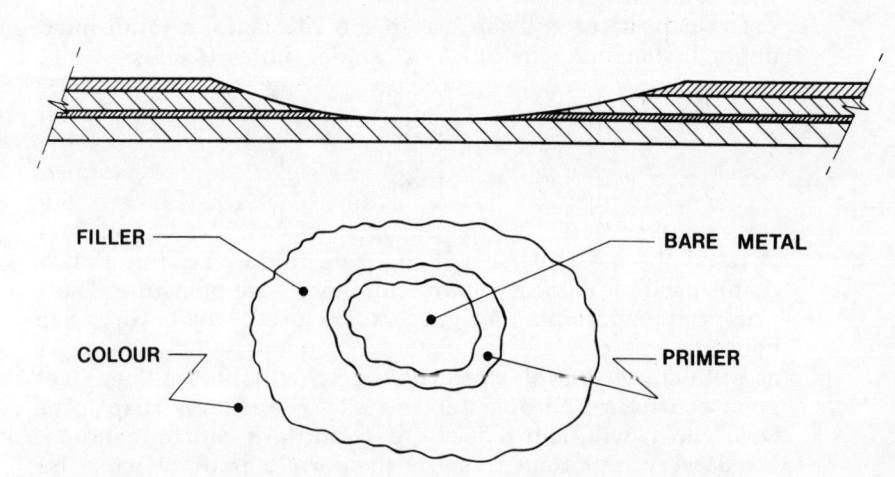

Fig. 17.   Feather edge

12.3 Wet flatting

When wet flatting, ensure that the correct grade of paper has been selected, and that the water is warm and clean. A sponge and good quality chamois leather are part of the tool list when wet flatting.

When flatting by hand ensure that:

A. The operator follows the contours of the panel.
B. The operator keeps his wrist well down to the panel.
C. The operator avoids finger 'furrows' and flats at right angles to the direction of the flatting action.
D. The operator washes the panel down continuously whilst carrying out the operation.
E. The operator changes the water frequently.

A rubber block is ideal for 'blocking' out over repaired areas and will ensure that the surface is flat. Too often these are not used and the result is that the hand tends to follow the irregularities of the repair.

The speed of the operation is important and as skill levels build it becomes easy to pace the operation so that a methodical and correct working speed is established. It is important to be effective and it is worth persevering for an extra fifteen minutes on a panel to ensure that it has received a total flat back. Often edges and areas around door handles (if they are not removed) are poorly flatted, which leads to paint flaking off, or if bridging has taken place, a large piece being torn away.

12.4 Dry flatting

Very much the same as wet flatting, if the operation is carried out by hand. If an orbital or flat bed sanding machine is used, then extra care is needed to ensure that the machine does not overcut. Digging in can cause problems later when rectification is necessary. Constantly check that the paper has not become clogged and therefore inefficient. If particles of paint get jammed between the surfaces bad scratching can occur. When the machine operation is complete, finish off on edges and returns by hand.

Clean the job well down using an intercoat solvent or spirit wipe to ensure that the lubricant is thoroughly removed and to show up any areas that have been missed during the total flatting operation.

Some refinishers prefer to start off dry and finish out by wet flatting which has merit, as the hard work is completed by machine and the final surface is finished by hand to give the totally effective job. The preparation of surfaces is an area in refinishing that should have great attention focused onto it. No

12.4 Vehicle Painter's Notes

matter how good the new technology paint systems are, if they are being applied over a poor and badly prepared substrate they will perform in a very much less reliable way. Attention to detail is vital to the final first class job. The difference is a painted vehicle that looks good and professionally painted and does not return under warranty, compared with perhaps a passable job, where the owner is back in six months with complaints.

Not only is it important to get a job out of the shop, it must remain out before being counted as a successful operation.

Getting the basics right is the foundation for good work, and correct preparation is the major basic.

13
What is Paint?

13.1 Definitions:
The definition of paint is 'colouring matter suspended in a liquid vehicle so as to impart colour to a surface'.

13.2 The components:
The basic components of paint are pigment, binder and solvent.

13.3 Pigment:
Pigments are finely ground powders derived from naturally occurring minerals or dyestuffs.

13.4 Binder or vehicle:
Binder or vehicle gives the film forming properties to paint.

13.5 Solvent or thinner:
Solvent or thinner 'thins' the paint for use.

13.6 Motor manufacturers' original finishes:
Various forms of primers and topcoats described.

13.7 Lacquers:
Acrylic lacquers dry by solvent evaporation and are impervious to ultra violet light.

13.8 Enamels:
The drying process and curing of high bake enamels.

13.9 Refinish paints:
Lacquer finishes – oil and synthetic finishes – low bake finishes – two-pack materials.

13.10 Refinish topcoats:
Nitrocellulose lacquers: These materials are a blend of cellulose and synthetic materials.
Acrylic lacquers: For use in the repair of high stoving acrylic finish. Very polishable and easy to use.
Air drying synthetic: High solid content material, often used to paint commercial vehicles.
Low bake enamels: The use of low bake finishes in the repair industry.
Two-pack acrylic enamels: A catalyst-cured high-build paint becoming very popular with the refinishing industry.

13.11 Practical differences:
The difference in refinishing materials.

13.1 Definitions

The full definition of paint is 'colouring matter suspended in a liquid vehicle so as to impart colour to a surface'. By this definition, paint is any coating from ladies' make-up to oil tanker deck paint.

Paint is usually described as any fluid coating that is applied and then dries out to form a hard continuous film. This is what is applied to modern motor vehicles and the material used by car repairers to repair that same continuous hard film.

A good refinisher should know all the undercoats and topcoats that are available and applicable for the repair of motor vehicles. That knowledge will enable him to choose the correct material and system to repair each job.

Refinish paints are complex and the paint manufacturers strive to improve the performance and reliability of their products and therefore offer the refinisher a choice of well-tested and approved refinish materials. The development of acrylic and two-pack materials, as well as the latest move in offering suitable coatings to the 'plastic' parts of modern vehicles, shows the great steps taken forward in paint technology.

13.2 The components

Automotive paints may vary in their properties and uses, but they all have three components in common: pigment, binder (vehicle), and solvent (thinner).

13.3 Pigment

Pigments used in the manufacture of paints are finely ground powders. These may be derived from naturally occurring minerals or they may be synthetic dyestuffs. Their properties are very important because they give the paint its hiding power or opacity, and colour, and help to determine its durability.

Pigmentation of paint depends on its functions. In primer, pigments help resist corrosion, in primer filler they are selected to give good build and easy flatting, and in finishes they give a lasting decorative effect.

13.4 Binder or vehicle

This gives the paint film forming properties, binding the particles of pigment together, and providing adhesion to the substrate.

13.5 Solvent or thinner

This makes the binder/pigment mixture fluid and workable during paint manufacture. It also reduces the paint to the correct consistency for application by spraygun, brush or other suitable method. The solvent mix is volatile, it evaporates both during application and after the paint has been applied, leaving the pigment and binder to form the hard continuous film. Proprietary blends of solvents are used to reduce paints to application viscosity. They are usually known as thinners. The term *reducer* can be used when the blend is specifically for synthetic finish enamel paint.

13.6 Motor manufacturers' original finishes

Four types of topcoat are used by the motor manufacturers and they are discussed in further detail later.

A. *Thermoplastic acrylic (TPA).* Used for both straight colour and metallic finishes. It will air dry, but is in fact stoved at 150–160° C (300–320° F) to give high gloss which results from a reflowing of the thermoplastic paint.

B. *High bake synthetic enamel.* Used for straight colour it has excellent filling properties, high gloss and toughness. It is unsuitable for metallic finishes as metallic control is difficult and gloss retention inferior to that of acrylic-based paints. It requires a temperature of 130° C (260° F) over a stoving period of 30 minutes for the full hardening process to take place.

C. *High bake acrylic enamel 'thermosetting acrylic' (TSA).* Used for both straight colour and metallic finishes. It also requires stoving for 30 minutes at 130°C (260°F) for the hardening process,

These three different paints cover the two basic types of paint used in both the motor industry and the refinishing industry.

D. *Two-pack urethane acrylic enamel.* Used for both straight and metallic colours, it may be either air dried or low baked at temperatures up to 80° C, which is 176° F. It is ideally suited for painting plastic components which could distort under the stoving schedules of high bake finishes. It is also used by commercial vehicle manufacturers where high stoving procedures are not practicable.

13.7 Lacquers

Acrylic lacquer dries and hardens by means of solvent evaporation only. It does not change chemically in any way and it remains soluble in lacquer solvents. When recoated with itself it is softened by the solvents of the repair paint. The two materials fuse into one and become bonded together. Such paints may be used for spot repairs and are known as lacquer paints.

13.8 Enamels

High bake synthetics and high bake thermosetting acrylics first dry by solvent evaporation and then harden by chemical reactions which take place at the baking schedule of 30 minutes at 130° C (260° F). The paints not only harden but become resistive to solvents. Any repair paint applied just lies on the surface, does not fuse or marry in and may well be wiped off whilst still wet without causing any harm to the underlying surface.

To enable high bake enamel finishes to be repaired after the car has been assembled and trimmed, and with the same original finish, a catalyst is added to the paint. This reduces the minimum stoving temperature to 90–100° C (194–212° F).

Acrylic urethane enamels dry and harden in a similar manner to high bake enamels with similar characteristics of durability and toughness. The reaction time is overnight at 20° C (68° F) or at stoving schedules ranging from 30 minutes at 60° C (140° F) to 15 minutes at 80° C (176° F).

A catalyst is not required and some motor manufacturers use them for 'in line' repairs of high bake finish.

Paints which harden by chemical reaction are known as enamel paints.

13.9 Refinish paints

Refinish paints may be classified into four types according to the drying process.

A. *Lacquer paints:* Drying occurs purely as a result of solvent evaporation and no chemical change is involved. Lacquer paints have a high solvent content and rapid initial dry.

This group includes: cellulose and acrylic vehicle paints, cellulose and acrylic-based quick drying primers, primer fillers, sealers and cellulose-based stoppers.

B. *Oil and synthetic resin-based paints:* Initial drying occurs by evaporation of the solvent, but that final hardening is due to chemical changes in the paint vehicle caused by the uptake of oxygen. Whilst these changes are taking place the vehicle gradually becomes less soluble in its own solvents and there is a critical period when lifting or shrivelling will occur on recoating. These paints are characterised by high solids content and a slower surface dry time than lacquer.

C. *Low bake enamels:* These paints will not air dry; to harden fully they need a baking temperature of 80° C (176° F) upwards. Initial drying is by solvent evaporation but final through hardening is dependent upon a chemical reaction between two components of the paint vehicle. This reaction can only take place in a time period of 30 minutes after the stoving temperature has reached 80° C (176° F). Low bake enamels have a high solids content. They also require the use of stoving equipment.

D. *Two-pack paints:* When the two component parts are mixed together a chemical reaction takes place causing the paint vehicle to harden. A drawback is the limited pot life or working time once the hardener catalyst is added to the material. Characteristics of the group are high solids content with low solvent content.

Typical finishes are: Two-pack polyurethane finishes, two-pack urethane acrylic finishes, two-pack spray primers and fillers, polyester stoppers and spraying fillers with peroxide catalysts.

13.10 Refinish topcoats

A. *Nitrocellulose lacquers.* These lacquers are a blend of nitrocellulose and synthetic resins. The nitrocellulose gives rapid drying. The synthetic resin content gives the high build gloss from the gun colour and gloss retention. Nitrocellulose lacquers dry by solvent evaporation only. As the initial rate of solvent evaporation is extremely rapid, so is the speed of the surface dry. However, some solvent is retained for a longer period. For full dry and through hardness 16 hours drying time is needed, in the proper conditions. They are: a temperature of between 68° F and 72° F with air movement. Conditions of cold and high humidity will affect the through dry, as solvent evaporating from the surface of the paint will not readily be taken up into the atmosphere. In poor conditions the solvent remains entrapped for very much longer periods and the film remains soft.

Nitrocellulose lacquers may be force dried or low baked at temperatures up to 70° C (160° F). Adequate flash-off time must be given to avoid popping.

When properly applied nitrocellulose lacquer will dry to a level film with a high gloss. If necessary polishing can be carried out to remove slight imperfections in the surface of the film. Polishing will also raise the gloss level.

The final appearance of the paint film depends on the use of the correct recommended thinner. It is essential to follow the manufacturers' specification on the use of thinners.

B. *Acrylic lacquers.* Acrylic lacquers are a blend of a resin from the acrylic family, such as poly methyl methacrylate (Perspex) and a synthetic plasticising resin. Acrylic lacquers very rapidly surface dry, and have excellent gloss and colour retention. Drying is by

13.10 Vehicle Painter's Notes

solvent evaporation only. This gives very rapid surface drying but for full through hardness 16 hours drying time is required in the correct conditions as mentioned for nitrocellulose. Acrylic is particularly sensitive to low temperatures. Force drying or low bake requirements are the same as those for nitrocellulose. Acrylic lacquers respond well to hand or machine polishing, using rubbing or polishing compounds. It is normally necessary to polish this finish to develop a proper gloss level.

Properly applied, an acrylic lacquer system can give excellent results. For durability, both the undercoat system and the thinners used must be those recommended by the paint manufacturer, and the system laid down must be followed. Failure to do this leads to failure of the paint film, usually by cracking or crazing up.

C. *Air dry synthetics.* Refinish air dry enamels are based on synthetic resins modified by drying oils such as linseed or soya bean during the manufacturing process. All enamels air dry in two stages:
1. The solvents or thinners evaporate from the film, which may then be handled and is dust free.
2. The paint film hardens by the absorption of oxygen from the air. After overnight drying it can be put into service.

Solids content of enamels is very high. Two coats will give a high gloss film with excellent scratch filling properties.

Solvents used in synthetic finishes are comparatively weak, so there is no danger of lifting or crazing of old paint films. This makes this material ideal for complete resprays of older vehicles as well as light commercials.

D. *Low bake enamels.* In low bake enamels, the paint vehicle consists of a blend of synthetic resins which will harden only when the paint film reaches a temperature of 80° C (176° F).

Stoving equipment must be installed to use this type of paint, which is similar in character to the high bake synthetic enamel factory finish used on many mass production cars. Application solids of low bake enamels are high. As with air dry synthetics, two coats will give a high gloss film with excellent scratch filling properties. Solvent mixtures are comparatively weak, and there is little danger of old paint films either lifting or crazing. Air drying synthetic enamels can be converted into 'low bake' enamels by thinning to spraying viscosity with special resin/thinner solutions generally known as 'hardener' or 'hardener/thinner'. In some cases the separate addition of hardener and thinner is necessary to reduce the enamel to spraying viscosity. Full 'low bake' temperatures are not necessary to harden converted enamels, but panel temperatures of 60° C (140° F) and upwards are recommended where high production rates are required. Typical stoving schedules would be 40 minutes at

60° C (140° F) or fifteen minutes at 80° C (176° F) metal temperature.

While hardness from the oven will not be quite to the standard obtained with a true low bake enamel, the motor vehicle may safely be put into service after stoving and the paint film will then reach full hardness after overnight drying.

E. *Two-pack acrylic enamels.* In high bake acrylic enamels, used extensively by manufacturers in the original finishing of motor vehicles, the paint vehicle consists of a blend of acrylic and melamine resins which only harden at a rate to meet production requirements when the paint film reaches a temperature of 125° C (257° F). Similar acrylic resins are suitable paint vehicles for refinish enamels when they are mixed, just before use, with a hardener based on an isocyanate resin adduct. They will harden at normal shop temperature, and the hardening process can be accelerated by force drying at temperatures of up to 80° C (176° F).

Two-pack refinish enamels have high application solids. Two coats will give a high gloss film with excellent scratch filling properties. They will cure at a faster rate than other refinish paints at any drying temperature from 20° C (68° F) to 80° C (176° F). Final hardness is equivalent to that of the high bake acrylic production finish. Recoat can be carried out, and petrol resistance is achieved after drying for 16 hours at 20° C (68° F), or at a typical force dry schedule of 30 minutes at a metal temperature of 60° C (100° F). After drying overnight at 20° C (68° F), or force drying, either as above, or at any other appropriate schedule, e.g. 15 minutes at 80° C (176° F), the film can be flatted and polished to remove dirt inclusions or even quite severe sags. The pot life, or working time, of these paints will generally be in the range of 4 to 8 hours. After this time all mixed paints should be discarded properly.

Other hardeners, free of the respiratory hazards of isocyanate, and giving longer pot life, are being developed.

13.11 Practical differences between the various refinish top coats

A. *Nitrocellulose lacquers:* These are tolerant of a wide range of application conditions. They surface dry rapidly and dirt in the work area is not a problem. Fast drying, along with easy polishing and good spot repair characteristics, give a high standard of finish even under conditions that are far from perfect.

The solids content of nitrocellulose lacquer is low. To obtain full gloss and good scratch filling of modern nitrocellulose synthetic resin finishes, high paint usage is necessary. Lacquer

13.11 Vehicle Painter's Notes

spraying thinners are strong and may cause old paint film to wrinkle, so great care must be taken when repairing older motor vehicles.

B. *Acrylic lacquer:* These surface dry slightly faster than 'glossy from the gun' nitrocellulose lacquer. Application is more sensitive to temperature change.

Polishing and spot repair characteristics are similar to nitrocellulose lacquer, and high paint usage is again necessary to obtain the required film thickness.

C. *Air dry synthetic enamel:* These materials dry to a full gloss in two wet on wet coats. Compared with lacquer paints the solids content is high therefore a synthetic paint is more economical to use. The weak synthetic thinners will not craze old paint films. Because of the comparatively long dust-free time required, drying conditions are critical. Application in a dust-free spraybooth is recommended to avoid dirt pick up.

The film will be soft and easily marked in the early stages of drying. There will always be a period, generally from 2 to 16 hours after the job is completed, when crazing may occur on recoating. Polishing is difficult, but after overnight dry and with skill and care, small blemishes may be removed by wet flatting with P 1200 paper, hand polishing lightly with liquid polish and/or machine burnishing at high speed, 6,000 rpm, using a dry lambswool mop.

D. *Low bake enamels:* True low bake materials which cure at a minimum panel temperature of 80° C (176° F) give fast, economical production, though at the cost of considerable investment in low bake plant. The paint film is hard and resistant to most forms of attacks, straight from the oven. No polishing is needed.

E. *Air dry synthetic enamel with hardener/thinner:* When an air dry synthetic enamel is used with a hardener/thinner, the refinisher gains the production advantages of a low bake oven without the need for the critical stoving schedules which true low bake enamels require. However, hardness from the oven is not quite so good. The paint film may be more easily damaged when the trim is fitted and some lifting may occur if areas with a high film build are recoated. Again, no polishing is needed but after cooling from the oven small defects may be removed by wet flatting with P 1200 paper, burnishing with rubbing compound (either by hand or machine) and finishing off with gentle pressure from a dry lambswool mop at high speed.

F. *Two-pack acrylic enamels:* Two-pack acrylic finishes will give more rapid hardening than other high solids refinish enamels at air dry and low stoving temperatures. The paint film is hard and resistant to most forms of attack after about 12 hours air dry, or after stoving for 30 minutes at a metal temperature of 60° C (140° F).

What is Paint? 13.11

No polishing is needed but after all night air dry, or immediately on cooling from the oven, the finishes may be brought up to a shiny mirror-like finish by wet flatting with P 1200 paper or burnishing with rubbing compound, finishing off with gentle pressure from a high speed (6,000 rpm) lambswool mop. These fast hardening times can only be obtained with paints having a limited working time of about 4 to 8 hours, once the two components are mixed together.

The foregoing shows how the refinisher must acquaint himself with all the various types of material on offer, so that he may choose wisely the preferred route when deciding what material to use on the work piece in hand.

The more background knowledge that a refinisher has of the products and their relative performances, the better the chances of correct procedures and therefore mistakes made can be limited and even totally eradicated by application of that knowledge.

14
Polishing

> **14.1 Object:**
> To cut back the colour surface to enhance the gloss of the final colour, to remove any dirt inclusions or debris or to remove road film prior to blending a colour coat.
>
> **14.2 Material:**
> The material is a cutting paste or compound, supplied in coarse, fine or extra fine, which both cuts and polishes. Can be hand or machine applied.
>
> **14.3 Use:**
> When hand applying use mutton cloth, damped, and worked on a small area at a time. Polish for lustre and finish off with liquid polisher such as T Cut to clear a milky appearance to the film. Avoid edges as break-through can occur, and use a lambswool bonnet mop for a brilliant finish.

14.1 Object

The object of polishing a panel or total vehicle is:

A. To cut back the colour surface to enhance the gloss.
B. To remove debris and dirt inclusions.
C. To remove road film prior to blending out a colour coat.

14.2 Materials

The cutting pastes or compounds are carefully formulated so that they will not harm the finish. They come in several grades from coarse to fine. They are very much like flatting papers as they in fact cut back the surface of the paint film. Always select the finest compound for doing the job, and do not use excessive amounts on thin final coats of colour as a break-through will occur, and that normally can be seen.

14.3 Use

When polishing by hand use a soft mutton cloth, clean and totally free of any contamination, slightly damped, and apply the compound sparingly. Apply gentle pressure and proceed as

Polishing 14.3

the compound cuts the surface. Check to see that excessive scratching is not taking place. Take care and avoid all edges, as these certainly will break through if any pressure is applied at this point. When compounding metallic finishes extra care must be taken to ensure that the surface is not broken, as that will immediately show up as a dark ring, and the only rectification is to respray the area concerned.

If the compounding operation has started with a coarse compound then finish off with a fine grade as this will remove the scratch marks of the coarse compound. Finish off with a liquid cutting polish such as T Cut for an extra fine finish.

When machine polishing, either compound or polish, it is an operation that must be carried out with great care. The machine can easily polish through edges, especially in awkward areas. Machine polishing can easily burn the finish and care must be taken not to apply too much pressure.

Polishing mops must be kept clean. Excessive compound and paint residues in a mop increase friction and the tendency to burn the finish. Lambswool mops can be washed with warm water and soap and then refitted on the machine and spun dry.

Before compounding and polishing a vehicle ensure that this operation is not carried out too soon after painting. Polishing of soft paintwork will lead to loss of gloss or hazing and the finish may be permanently impaired.

The application of wax polish too soon will cause water marks and softness. In an ideal situation it should not be applied until the vehicle is at least four weeks old. The wax can migrate into the paint film and cause loss of gloss, therefore it should be avoided.

Great care should be exercised when compounding and polishing a vehicle, as the final finish is all that the customer sees. The body repair underneath may be good but he cannot see this and only the very top surface of the paint film is actually seen.

Great care should be taken at every stage of painting a vehicle. If this is done then the use of compounds and polishes can be kept to the very minimum, which is much more preferable in any case, as the more a film is pulled about the less effective it becomes as a total film. Paint chemists will always prefer the refinisher to obtain a satisfactory gloss from the gun and not touch the final paint surface at all. It is a good point to keep in mind.

15
Safety

> There are certain key hazards in the refinishing industry.
>   Fire and explosion are the most important, followed by the hazards presented by the incorrect use of compressed air, or the failure to use safety and breathing equipment whilst working with spray materials and acid-based preparations.
>   Safety is the responsibility of everyone present in a working environment.

The need for safety at work cannot be stressed enough. Many workplaces are hazardous but with just a little care and thought the hazardous conditions could be improved to create a safe place of work. The refinishing industry, fragmented as it is in the UK, has grown up round the backs of garages, down little alleyways, and out of the town. Nowadays, main agents who have started up refinishing shops have done the thing properly and taken expert advice. Safe shops have proper facilities for preparing and spraying vehicles. Paint stores are positioned correctly, away from the main building, and there is adequate care of personnel using the materials.

Without doubt fire remains the greatest hazard. Everything that can possibly be done must be done to ensure that fire cannot break out, but if it does it must be contained quickly and effectively. Fire extinguishers and fire blankets must be on hand and in a working condition.

Safety equipment such as masks and air-fed face masks must be on hand and used when flatting polyesters and spraying all paints, with special emphasis on polyisocyanates. Many refinishers do not fully understand the long-term risks attached to using these materials if they do not follow the safety recommendations.

Handling motor vehicles can be dangerous if they are driven around the shop. Cars can go out of control and staff can be run over. Vehicles can and often have been driven over operators' feet, with disastrous long-term consequences.

The need for safety and safe operations is paramount. It is encumbent on every person employed to behave in a safe manner and have the care and safety of his fellow worker at heart.

16
Removal and Storage of Parts

16.1 Object:
To remove all body trims and accessories to obtain a properly painted vehicle, and to attend to rust areas behind trim and motif attachments.

16.2 Removal:
Exterior chrome mouldings and motif badges are often a press fit, some may be adhesive backed. Always work with care and remove items at right angles to the panel.
All items should be stored in the vehicle or in numbered bins.

16.1 Object

Often the paintshop will be involved in removing small parts from vehicles prior to painting. Depending on the shop operation this function can be shared between the bodyshop and the paintshop. The end result is the same, and that is that as many parts as possible should be removed prior to painting. Masking up of small items such as badges or motifs can take time and no matter how carefully done always seems to show some overspray or edge. When a motif is removed and the paint is sprayed over the entire panel a much better job is the result. The cleaned motif may be replaced giving a much better and finished look to the job.

Often rust has commenced behind strip finishers or badges and it is necessary to remove them to attend to the rust contamination.

16.2 Removal

Exterior chrome finishers and motifs are normally a press fit. Usually the fasteners are of plastic to contain the rusting problem associated with the old metal clips. Many badge and type motifs are adhesive-backed, so these have to be replaced with new pieces once they have been removed.

Always remove parts with great care and at right angles to the vehicle or panel.

Normally lights and bumpers are fixed from inside the vehicle which generally ensures that any rust attack has been minimised.

16.2 Vehicle Painter's Notes

All items removed should be stored in a bin that is allocated to that vehicle but often the boot of the vehicle is used as the store.

Do ensure that all items are cleaned before refitting. To attempt to clean after fitting is unsatisfactory and always marks the paintwork.

17
Power Tools

17.1 Types:
Air or electric – DA sanders – random – orbital. For safety – air tools preferable.
Orbital sander – block shape incorporating movable rectangular head – covered by abrasive paper.
Random orbital – discs glued to pad – circular motion in orbit around a circle.

17.2 Operation:
M/C balanced – little pressure needed – best finish use tool lightly – allow to 'float' over surface. Place sander on work before starting and lift before stopping – speed adjusted by air pressure.
Feather edge – tilt tool 5° and finish flat.
Keep M/C clean and oiled.

17.1 Types

The range of power tools currently on offer is indeed large and all-embracing. They are either powered by compressed air or electricity and for safety air-powered tools are preferable in the paint shop.

The sanders are either orbital or random orbital.

A. An orbital sander is a block-shaped unit incorporating a movable rectangular head. This is covered by abrasive paper and that is held by clips on the reverse or upside of the block. The air tool can of course be used with wet and dry paper in wet form.

B. A random orbital sander is a disc pad machine that has a circular motion in orbit around a circle. This unit is generally used in the dry form with Stick It discs of various grits.

17.2 Operation

The machines are very well balanced and need little pressure to operate very effectively. To get the best finish, use the tool lightly and allow it to 'float' over the surface.

Always place the sander on the panel before starting it and lift it off before stopping.

17.2 Vehicle Painter's Notes

The speed of the tool can be adjusted by the air regulator fixed in the body of the unit.

The tools are ideal for feather edging and can save hours of work when used correctly.

Never tilt the tool more than 5° from the horizontal and try to keep the tool face flat as you follow the contours of the panel.

Keep the machines clean and free from dust. Oil them daily or better still, fit an oil dispenser between the air line and the tool to dispense oil during the operation.

18
Fibreglass

18.1 Material:
Glass reinforced polyester is being used more and more in the automotive industry for parts such as air dams and spoilers. Lotus and Reliant have made great contributions to the development of the material.

18.2 Procedure:
To prepare parts and panels ready to accept a full paint system and then followed by the full painting procedure.

18.3 Faults:
A list of faults that can occur and the rectification at every stage.

18.4 Low bake:
Special lower temperature is required to use low bake materials successfully.

18.1 Material

The use of glass reinforced polyester has increased dramatically over the last decade. More and more mass-produced cars are using GRP in one form or another. The whole range of 'plastics' is becoming standard use within the motor industry, for all body panels and parts from spoilers and air dams through to bonnet and boot lids. The famous GRP vehicles such as Lotus and Reliant have done much to promote the use of these new materials.

The structure of both resins and matt has improved and the standard and consistency have made the necessary impact on designers. Not only the specialist manufacturers such as the above-mentioned Lotus and Reliant, but also TVR and even Rolls-Royce and Aston Martin Lagonda Ltd are using this material in ever-increasing amounts.

In the commercial vehicles, coaches are using GRP in cab construction and front panel mouldings. The material is lighter and 5 times as strong as steel at the same thickness.

The repair of this material has become relatively straightforward with the use of improved resin technology. The refinishing after repair offers no difficulty provided the sequence is followed carefully.

18.2 Procedure

If a new panel has been fitted to a vehicle, such as a wing, then it is absolutely necessary to remove all trace of the wax release agent used when the panel was moulded. This is done by washing the panel down in very hot soapy water and wet flatting using more hot soapy water. The following sequence should be followed:

A. Wash down with hot water containing a liquid soap.
B. Wet flat using P 400 paper.
C. Wipe off with spirit wipe.
D. Dry off and examine for imperfections and pinholes.
E. Fill up pin holes with catalyst body filler.
F. Rub down any imperfections.
G. Wipe over with spirit wipe.
H. Apply a single coat of self etch primer.
I. Apply two or three coats of filler.
J. Apply guide coat and wet flat accordingly.
K. Apply colour in the normal way.
L. Polish finished job as normal.

18.3 Faults

It is important to follow the procedure outlined above to ensure a first-class job. However, faults can arise from the gel coat of the GRP and these need to be rectified.

If air pockets are trapped in the gel coat they are likely to:

A. Break through when flatting back the surface during initial preparation.
B. Break through during a low bake process.

In A., action can be taken to rectify at the early stage when filling can take place, but in B., the problem is much more difficult. Stripping back to the gel coat surface is the only certain way to overcome this problem. Any other filling or stopping is only an uncertain attempt at curing the fault.

Often after an accident GRP suffers stress point failure which manifests itself as 'starring'. Some stress forms up in lines and both these faults are shown in Fig.18. Rectification for this fault is as follows:

A. Clean back any paint from the surface of the gel coat.
B. At the end of each stress line drill a 1/16 inch hole.
C. Groove up the stress line with production paper.
D. Body fill the groove and the hole at the end with catalyst bodyfiller.
E. Rub the repair down with 80 grit production paper.
F. Wet flat with P 400 paper and proceed with the paint system.

Fibreglass 18.4

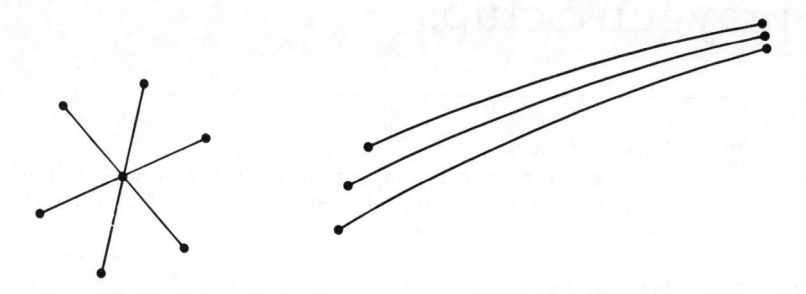

Fig. 18. Fibreglass stress points

18.4 Low bake

It is recommended that when low baking GRP a panel temperature of 60° C is the maximum. This rules out the use of the old low bake materials that required a panel temperature of 80° C to properly cross link. A lower temperature means that paint can either be force dried or, in the case of two-pack materials, the catalyst action takes place more quickly due to the introduction of the heat.

Lotus have been using the ICI 2K two-pack material very successfully on their GRP shells. The finish obtained is very good indeed.

19
Spraygun Setups

19.1 Varying factors:
To obtain the correct spraying pressure the varying factors such as the type of paint, cellulose or acrylic type of thinners, fast or slow, must be taken into account.

19.2 Pressure too high:
This will cause excessive paint loss and poor flow. During a full respray this fault can cause excessive dry spray.

19.3 Pressure too low:
This will produce a paint film with poor dry through and will be prone to sagging and solvent retention, which causes a soft film.

19.4 Atomisation:
To check the correct atomisation of the paint and the spraying pressure spray out material on a paper sheet attached to the spray booth wall, and adjust the gun controls.

19.5 High pressure faults:
A full list of faults experienced when the pressure is set too high.

19.6 Low pressure faults:
A list of low pressure faults.

19.1 Varying factors

The setting of the correct spraying pressure revolves around a number of varying factors. The discipline and attention to the detail of the job ensures that these factors are properly attended to, and the resultant spraying application is correct. These factors must be correctly determined before the spraying commences, and time spent at this stage is well rewarded. The correct spray pressure will depend on a combination of varying factors:

A. The type of paint to be used, e.g nitrocellulose, acrylic, synthetic or two-pack materials.
B. The type of thinner to be used, e.g. fast, slow, retarded.
C. The viscosity of the material.

D. The air cap capacity.
E. Paint emission rate.
F. Pressure gauge accuracy.
G. Pressure drop through the air line.

19.2 Pressure too high

The optimum spraying pressure is the lowest needed to obtain proper atomisation, emission rate and full fan width. Too high a pressure results in excessive paint loss through overspray and therefore high paint usage, and in poor flow due to high solvent evaporation before the paint reaches the surface being sprayed. When a full respray is being attempted the condition of high pressure can be disastrous as the dry overspray settles on the vehicle and is impossible to wet up. Often a full wet flat is the only course open to the operator.

19.3 Pressure too low

Too low a pressure will produce a paint film with poor drying characteristics, due to high solvent retention, and the film will be prone to bubbling and sagging. In low bake this will cause the solvent to pop in the paint film, or solvent boil.

Before painting the spray pattern should be checked to confirm the proper atomisation and size and shape of the spray pattern. Often operators will just spray the gun into the air and look at the fan. This is not accurate and it is worth taking trouble to get it right.

There is a very simple check system to determine the optimum spraying pressure for any paint, at any reasonable viscosity with any spraygun.

Sometimes spraying pressures may be either higher or lower than the optimum because of special circumstances such as the spraying of metallics for certain effects, or special fast or slow thinners or even adverse climatic conditions. This does not apply in a properly controlled spray booth, however.

19.4 Atomisation

To check correct atomisation and spraying pressure:

A. Attach a sheet of paper to the wall in the spray booth.
B. Adjust the air pressure at the regulator to approximately 30 psi (2 bars).
C. Adjust the spreader and fluid needle controls on the spray gun to fully open position.
D. Hold the gun a set distance from the paper surface and spray for 2 to 3 seconds keeping the gun in the same position. An

easy way to set the correct distance, which should be 6 to 8 inches, is to use the distance between the thumb tip and the little finger tip of the outspread hand.
E. Raise the spraying pressure by 5 psi (0.3 bars) and repeat the static spray operation next to the first spray.
F. Continue to repeat this operation, raising the pressure by 5 psi (0.3 bars) each time until the pressure that gives the maximum spray pattern is established.
G. Attach a clean piece of paper to the booth wall and keeping the gun perpendicular and the same distance from the paper surface, make a very fast pass across the paper with the trigger right back, the gun fully open. The gun must pass across the paper fast enough to allow the paint particles to fall separately on the surface of the paper.
H. Raise the spraying pressure by 5 psi (0.3 bars) and repeat the procedure. Then compare the paint particle size. Then continue to repeat raising the pressure each time by the 5 psi (0.3 bars) as long as necessary to determine the pressure which gives the finest or at least adequate atomisation. See Fig.19.

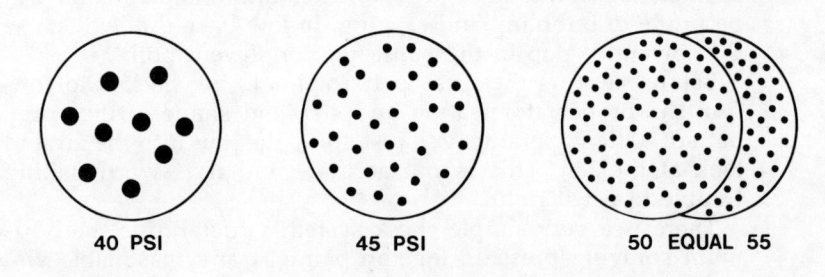

Fig. 19. Spray atomisation – correct spraying pressure is 50 psi

19.5 High pressure faults

The correct spraying pressure is fundamental to the successful spray job and it is essential to do it right.

Air pressure that is too high will lead to:

A. Cobwebbing of the paint as it leaves the gun.
B. Dry spray over the job and in the environment.
C. Orange peel finish.
D. Low gloss due to over-atomisation.

19.6 Low pressure faults

Air pressure that is too low will lead to:

A. Floating of the colour.
B. Popping of the solvent trapped in the film build.
C. Pinholing due to solvent trapped.
D. Runs and sags in the final coats.

To ensure that the requirements of both volume and pressure of the gun are fully maintained, it is essential to take account of air tools and guns running off the same compressor.

20
Machine Polishing

> **20.1 Object:**
> Rapid polishing of a panel or vehicle.
>
> **20.2 Uses:**
> New painted panels of poor gloss or second hand car refinishing – areas etc. Lambswool bonnet fitted to a high-speed air driver or electric machine, compound applied, followed by T Cut, 'mopped' to high gloss.
>
> **20.3 Disadvantages:**
> Can bruise paint if too fast – cut through edges, cause paint to go flat if polished too soon after painting.

20.1 Object

Machine polishing is designed to give a rapid buff to a finished painted surface.

20.2 Uses

It is used on freshly painted panels of poor gloss levels or for resurfacing a second-hand car. Sometimes it is used as part of a polishing and cutting procedure for the ultimate gloss level required by a specialist motor manufacturer.

Aston Martin Lagonda Ltd, who present their vehicles painted to the highest possible standard, in fact wet flat the final colour or lacquer coats and machine polish to raise the gloss level to an extreme and then finish off by hand with very fine compounds and polishes. The result is outstanding.

A lambswool bonnet is fitted to a high-speed air- or electrically-driven polishing machine. A compound is applied to the panel and the mop is then set in motion to cut back the surface using the lambswool mop. Eventually the compound is cleared from the surface and the mop burnishes the surface.

20.3 Disadvantages

If the machine is running too fast or too much pressure is put behind the mop then burning of the paint film can occur. Edges

Machine Polishing 20.3

can easily be cut through exposing primer coats underneath, and in the case of metallic finishes the polishing can cut through the lay of the metallic giving a silver or dark ring around the area that has been breached.

If machine polishing is carried out too soon after painting it can cause the finish to go flat. Leave at least three days after painting before machine polishing.

21
Spray Booths

> **21.1 Essential equipment:**
> Use and basic requirements for a normal air dry spray booth must be that it is operating correctly with full filtration of incoming air and it must be heated. Proper flame-proof lighting must also be in operation.
>
> **21.2 Requirements:**
> The booth may be pressurised to exclude dirt, and it should always be run before spraying so that the temperature may be correctly set and any dirt or dust has settled or been drawn out.

21.1 Essential equipment

It is essential that, no matter how small a refinisher is, he has on hand a suitable spray booth to cope with his refinishing requirements. For safety and the reduction of dirt inclusions a well-operated spray booth is necessary for basic work.

Any spray booth will function effectively provided the following basic requirements are met:

A. Incoming air is properly filtered.
B. Airflow is in the direction of the pull of gravity, in other words from the ceiling to the floor.
C. The air velocity is in the region of 50–120 linear feet per minute (16–40 metres) and the air volume is such that a minimum of two air changes per minute is achieved.
D. The extraction is uniform, that is the air flow gives an even fume extraction all around a vehicle situated centrally in the booth.

21.2 Requirements

The booth may be pressurised with air forced in at a slightly greater rate than it is withdrawn. This creates a positive pressure within the booth which prevents any dirt entering from outside. The booth should always be run before painting the first vehicle to allow clean spraying conditions to become established. In an unpressurised booth it is essential that all doors are sealed 100% effectively to prevent dirt being drawn in when the extractor fans are running. Over-pressurisation should be avoided to prevent the discharge of paint fumes into the open shop.

The extract air may be drawn over water or through a water curtain beneath the spray booth floor. This reduces the fire hazard and prevents the discharge of spray dust from the booth.

22
Accurate Masking

> Always ensure that the vehicle is clean before masking up. Use a small tape as the starter and leave a minute gap between tape and the surface to be painted to avoid bridging.
> Fill in accurately and do not stretch the tape.
> Take time and care as accurate masking will save hours of rectification work.

It has often been said that if you could only mask up a vehicle as quickly as you unmask it then life would be very easy and a lot less tedious. Well that may be so, but it is a fact that very little time is devoted to very careful and accurate masking of a panel or whole vehicle. For a good finished job it is essential to ensure the highest standard of masking.

Faults that occur through bad or poor masking are numerous. Bridging is the first and potentially the most damaging. If the tape and paper are pulled away with the paint film bridged to

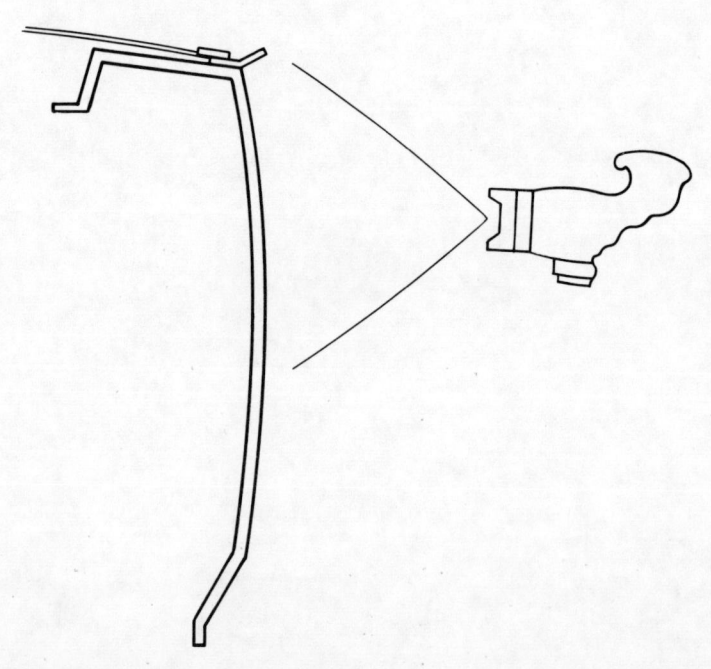

Fig. 20. Masking wing edge and top

Accurate Masking 22.1

it the resultant tear in the paint film can cause great difficulty in spotting in, particularly if it is a metallic finish or a clear over base. Often this simple fault may necessitate the respray of a complete panel.

Contamination of the paint finish can result through not using the proper masking paper, which can in turn call for a repaint to rectify.

When masking accurately ensure that no undue stretching of the tape takes place as this is likely to buckle up when being stoved in a low bake oven.

Using tape correctly enables the operator to paint up to a body line, such as a wing top edge, and then complete an invisible repair. Metallic finishes can be repaired in this fashion where by using careful masking techniques lower panels can be painted without coming up with finish into an 'A' surface area. In Fig.20 the top of the wing is masked and the lower panel painted. After removal of the masking the edge of the wing can be lightly compounded so that the join cannot be detected.

Develop accurate and clever masking to enhance the finished piece.

23
Vehicle Wiring

23.1 Object

To carry out current to all parts of vehicle and supply power to operate electrical items.

23.2 Polarity

Means to describe the way a battery is connected – positive or negative earth to car body.
Diodes (valves) and transistors (switch) are polarity-conscious – current flowing in opposite direction will damage.

23.3 Circuit

Battery – fuse – switch circuit – lamp – colour coded – wiring diagram has to be obtained.

23.4 Failures

When lamps fail to work on refit, the normal cause is bad earth – secondary cause is lamp failure.
Remember to look – think – act.

24
Spray Booth Coating

> **24.1 Description:**
> This material is a tough, elastic, self-supporting film, resistant to many substances, that does not support combustion.
>
> **24.2 Use:**
> In all spray booths to give both good illumination and cleanliness. It is a ready-for-use material and should be applied by spray.

24.1 Description

It is essential for good housekeeping and maintenance to ensure that the spray booth is well lit and clean. It is most advisable to paint the surface of the booth with an interior booth coating. This material can be readily supplied and is essential for absolute cleanliness.

The material is a very tough, elastic self-supporting film. It adheres well to the booth interior but when it has become contaminated with overspray it can simply be peeled off the booth wall.

It is fully resistant to oil, petrol, alcohols, fats, acids and alkalies. It is designed for interior use only. It will not support combustion and if a fire does occur then when the film is heated fire-retardant fumes are given off.

24.2 Use

For use in all spray booths, it will give good reflective illumination. No preparation is necessary, just spray the material at 60 lb pressure and apply two coats. This will give a self-supporting film of about two thou build. It is dry in about ten minutes and the booth can be put back to work. 5 litres of material will give 165 to 220 square feet coverage.

25
Finishes

There are four types of drying or curing systems for vehicle finishing and refinishing material.

25.1 Solvent evaporation:
These finishes are usually cellulose or acrylic and have rapid surface dry and are very polishable. They are easy to recoat.

25.2 Oxidation:
These materials are synthetic, both quick dry and full 16 hour or overnight dry. They take up oxygen from the air and polymerisation takes place.

25.3 Low bake:
These finishes cure by the application of heat. The panel temperature required for the material to cross link is 80° C.

25.4 Catalyst materials:
These materials cure by catalyst action. They can be air dried or force dried to speed the cure out.

The finishing and refinishing materials that are currently used throughout the automotive industry fall in to four broad categories for drying out.

25.1 Solvent evaporation

Materials that dry by solvent evaporation from the total film. These materials are most commonly used by small refinishers for accident repair work. At one stage they were the most widely used materials but more modern advances have dated the performance of these finishes.

Generally the materials are nitrocellulose and acrylic finishes with their substrate primers and surfacers. They have a rapid dry after application provided that the air temperature is between 68 and 72° F. They are very polishable and can be flatted easily and reworked without difficulty. Dirt inclusions and nibs can be easily flatted out and the surface polished or resprayed. As these materials dry by solvent evaporation no chemical change takes place and the material just sinks back into the substrate below. The need for a good primed and filled undercoat system is of great importance as the drying out and consequent sinkage will show up all imperfections and 'map' the surface. The main advantage is the ease with which it can be recoated and represented without any undue problem.

25.2 Oxidation

Materials that dry by oxidation and the film is polymerised as a result. The paint film is applied to the surface and then after flowing out begins to take up oxygen from the air and harden. Synthetic and quick drying synthetic finishes come into this category and are very popular for commercial vehicle finishes.

As the paint film cures from the outside through to the substrate there is what is known as a critical recoat period. This means that after application of the final finish there is a refinish time of up to three hours where further coats could be applied. After this time has elapsed and the film has started to cure, to recoat for any reason would be impossible without some reaction taking place, until at least 48 hours had elapsed from the final colour application. Always consult the manufacturers' recommendations when using this type of material.

The advantages of these oxidising finishes is the very high film build that they offer. With up to 50 per cent solids the film build and covering power is very impressive. Once fully oxidised, the gloss and durability of this material are very extensive and the material is very hard wearing. Because of the high build, any imperfections in the substrate are filled by the film build.

25.3 Low bake

Materials that cross-link when exposed to heat are the low bake finishes, and in the case of original finish at the motor manufacturers, a high stoving finish. Both low bake and high bake dry initially by solvent evaporation that allows the film to 'set up' and then as heat is applied the molecules of the system cross-link and fuse together. In the case of low bake this takes place at a panel temperature of 176° F or 80° C. High stoving cross-links at 320° F, which is 170° C. As the panel cools to ambient it takes up oxygen to complete the hardness of the film, as all low bakes when hot tend to still be thermoplastic. These original finishes give excellent gloss and long-term weatherability and can be repaired with ease.

25.4 Catalyst materials

The final group of materials are those that dry out by catalyst action. These are known as two-pack materials and there has

25.4 Vehicle Painter's Notes

been a most significant growth in their use within the UK during the last ten years. The base material is a polyurethane high build that offers massive opacity levels and high gloss, and the hardener or additive is an isocyanate material. This is mixed with the base and then it is thinned for spraying. It gives a remarkable build, with the drying time of a lacquer that dries by solvent evaporation. This of course allows the refinisher to have a high throughput with material that will cover any blemish in the substrate and give an excellent gloss with good and lasting weatherability. This material can be force dried as well as being left to air dry.

It is important to show the difference between force dry and proper low bake. It is possible to force dry any material. The application of heat speeds up the drying process that would have taken place in any case at the temperature window of 68 to 72° F. Cross-linking does not occur in this case. The greater the temperature application, the faster the cure through. Low bake, however, will never properly cross-link until the application of heat at the prescribed temperature. They will appear through dry after they have been left for some time but they are not fully cured in the true sense.

In the case of spraying materials that dry by isocyanate it is essential that full breathing equipment be used. It is good practice to always use full breathing apparatus when carrying out any spraying.

26
Rust Spots

> 26.1 **Rust spots:**
> Normally a stone chip or blemish occurring in the paint substrate will cause the rusting process to commence.
>
> 26.2 **Procedure:**
> The full procedure to repair the rust spot effectively.

26.1 Rust spots

Sometimes due either to a stone chip or a blemish occurring in the substrate, a rust spot will appear in the film. If this spot is due to rusting out from the metal panel then the only recourse is the repair of the steel, usually by welding. However, if the problem is a local one then the following procedure should be followed.

26.2 Procedure

A. Compound the area and remove all road film.
B. Either wet flat with 180 wet or dry, or using a DA sander with a 240 grit disc, flat out the spot and roughly feather edge.
C. Examine the area carefully to ensure that the rust contamination has been fully cleaned away from the surface of the metal.
D. Then wet flat a good feather edge around the broken paint film.
E. Dry off and wipe over with spirit wipe or wax and grease remover.
F. Apply a primer coat as per the manufacturers' instructions.
G. Apply coats of filler to the area.
H. When through dry, guide coat and block back with P 600 paper.
I. When well flatted clean down and wipe over with spirit wipe.
J. Apply three coats of colour thinned to paint manufacturers' recommendations.
K. Apply a thinner coat to the edge of the 'blow in' to soften down the edges.
L. Allow to through dry and either compound or wet flat with P 1200 and compound and polish.

27
Vehicle Manufacturers' Materials

> Vehicle producers in Europe have standardised systems and they generally follow the same technical route.
>
> Ford:
> Prime up the steel monocoque body in electrophoretic primer followed by acrylic top coat stoved at 280 to 320° F for 30 minutes.
>
> G.M. Vauxhall:
> Prime up the steel monocoque body in electrophoretic primer followed by an acrylic colour coat which is stoved and then lightly flatted and restoved at 280 to 320° F to obtain a 'reflowed' finish without dirt inclusions or runs.
>
> Austin Rover:
> Prime up the steel monocoque body in electrophoretic primer followed by acrylic colour coat which is stoved and lightly flatted and restoved to obtain 'reflow' finish.

Motor manufacturers in the UK and Europe do have some standardised systems of painting motor vehicles.

The object of a manufacturer's system is to paint a monocoque shell to an even depth with the best weatherability factor and gloss level in the shortest possible time. Paint technology has been advanced to such a stage that within the parameters set by the motor manufacturers the materials of today are of extremely high quality, especially within the realms of durability. To add to the consistency of performance, robot application of paint has had a significant effect on speed as well as conformity of total film build.

Ford Motor Company have influenced the market in many ways and certainly in the field of original painting they have often led the way. The introduction of electrophoretic primer, which is the deposition of primer to the steel shell, was a major breakthrough in primer application with such good corrosion resistant qualities. By electrodeposition primer was drawn into the most confined parts, such as between welded seams, which gave protection that normal dipping procedures simply could not. Also, depending on the current supplied, an accurate film build could be controlled. Normally the film thickness is between 1 and $1\frac{1}{2}$ thou (25 microns to 37.5 microns) and this is ample to give the protection desired, and this film is completely uniform. Simply put, as the film build increases it acts as an insulator against the electrical charge so that only when the primer reaches the correct build all over does the process stop. The paint primer system is used by all the majors.

Vehicle Manufacturers' Materials 27.1

In the late '50s the American canning industry first looked at acrylic finishes. The advantages of this water white resin were quickly developed up for the automotive field. Acrylic resin is a plastic within the same family as perspex and PVC (poly vinyl chloride) and it was developed so that colour could be built into this material that is impervious to ultra violet light. The American automotive industry used this as original equipment finishing and it naturally was later used in the UK and Europe. The manufacturers use it in two forms. Ford use it as a single paint and bake system (thermosetting acrylic) and General Motors Vauxhall use a single paint followed by bake then sand and reflow bake. Jaguar Cars also use this system, known as thermoplastic acrylic or TPA. The advantage of this system is that after the vehicle unit is first painted and stoved, any runs, dirt inclusions or orange peel can be flatted out before the reflow takes place. As high temperature is applied (320° F) the flatted colour 'wets up' and the gloss level returns. This system has done much to enhance the quality of the paintwork on new vehicles. Austin Rover have now moved on to TPA with the resultant improvement to finish, and as the motor industry has now realised the expectation levels of the purchasing public, greater emphasis has been placed on the appearance of the vehicle at the point of sale.

Now, in an attempt by the market strategy teams to attract more sales to the over-production that has been geared up for a shrinking market, the new pearl finishes made their debut in the UK during 1986. Jaguar are leading the way and others will surely follow. The introduction of clear over base on the line is another significant move, and this coupled with the pearl effect colours will tax the refinisher even more. The operator in a busy repair sprayshop, or main dealer, is going to have to learn new painting techniques. They are not difficult but they must be understood and handled accordingly. Discipline of procedure and full understanding of what is involved will make the painter's job easier. Foremen and managers will have to understand the difficulties and problems associated with refinishing these types of finish.

To repair all of the types of finish outlined is not too difficult for an experienced and careful painter. The normal total film builds on new vehicles range from Ford at 3 to 5 thou (75 microns to 125 microns) to Jaguar at 6 to 8 thou (150 microns to 200 microns) and these finishes correctly feathered out will take any form of material currently on offer to the refinishing industry. Colour matching and especially metallics and pearl finishes are dealt with later, but following basic rules it can be well accomplished.

It is unlikely that the painting of vehicles by the manufacturers will alter much in the foreseeable future, even with the certain introduction of 'plastics' to the line assembly pro-

27.1 Vehicle Painter's Notes

cedure. New developments in colour and long-term durability will be the most likely future route. Volvo have announced an eight-year warranty on paint and body, and others will follow.

The painting of 'plastics' and 'soft parts' is dealt with later but it is interesting to note that there is expected to be a most dramatic rise in the use of these materials in modern mass-production vehicles.

28
The BIA Research Centre at Thatcham

> **28.1 Background:**
> The BIA was set up in 1970 to study repair methods and times on motor vehicles.
>
> **28.2 Designing for repairability:**
> Good repairability means that, for a measurable standard impact the total repair cost is low.
>
> **28.3 Designing for damageability:**
> Good damageability means that, for a given measurable impact, the extent of the damage is low.
>
> **28.4 The average accident:**
> The average accident damage is outlined and shown by diagram.
>
> **28.5 Thatcham times:**
> Times needed or required to repair and refinish vehicles.
>
> **28.6 Key to success:**
> The key to success in repairing vehicles is to have a well equipped workshop employing professional staff.
>
> **28.7 Training:**
> The training of personnel is vital if a paint and bodyshop is to remain competitive and effective.

28.1 Background

The British Insurance Association (BIA) and Lloyds set up a research centre at Thatcham in Berkshire to study the repair procedures and costs of repairing damaged motor vehicles. The centre opened in 1970 and has grown dramatically since those early days. It was set up to produce realistic times for panel replacement and final painting.

The cost of repairing vehicles in the UK has now risen to £1.3b annually.

The BIA now work closely with the motor manufacturers at the design stage, advising the manufacturers as to the repairability of the model. This obviously will affect the insurance rating of the vehicle, and the manufacturers are anxious to give the customer the benefit of a low insurance classification.

28.2 Designing for repairability

Good repairability means that for a measurable standard impact situation the total repair cost is low. Low cost can either be due to low parts cost or low labour hours or a combination of the two. A vehicle must be 'repairer friendly', providing good accessibility. Good repairability does not equal good damageability. It is all very well claiming easy and cheap repair characteristics but the question must be posed – should it have been damaged in the first place?

28.3 Designing for damageability

Good damageability means that, for a given measurable impact situation, the extent of the damage is low. This necessitates good energy absorbing characteristics, either by progressive collapse of the structure or by the use of energy absorption materials or both.

28.4 The average accident

The average damage distribution in the UK is set out in Fig.21. For the rest of Europe it is much the same. However, Fig.22

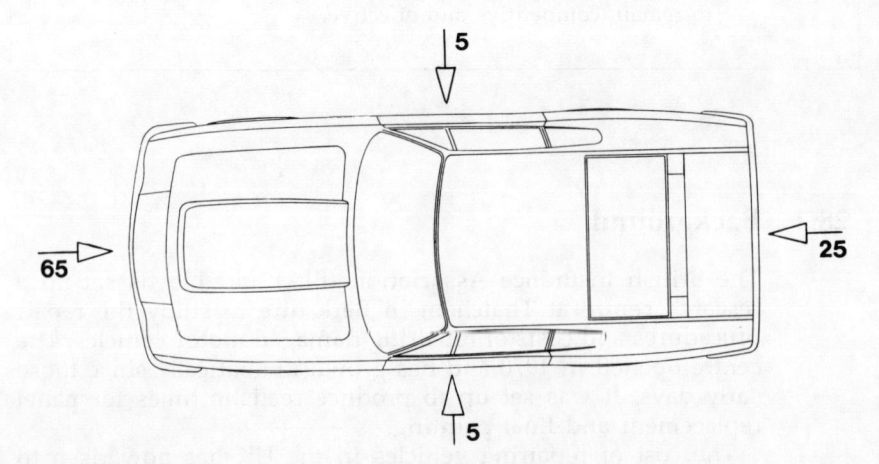

Fig. 21. UK damage distribution (percentage)

shows the 40% offset damage which equates to the highest proportion of accidents. In a busy repair shop with a painting facility it can be expected that a large proportion of the work will fall into these parameters. This means that total resprays are indeed rare, and therefore colour and original finish surface matching play a large part in the time taken to repair the vehicles.

The amount of material used in a paint shop is an important criterion with paint and materials now becoming more and more expensive. Figure 23 shows a table produced by the BIA for unthinned materials used on the various panels, listed by vehicle model. These amounts are only a guide but they are useful in helping to promote some control and consistency in the amount of paint consumed.

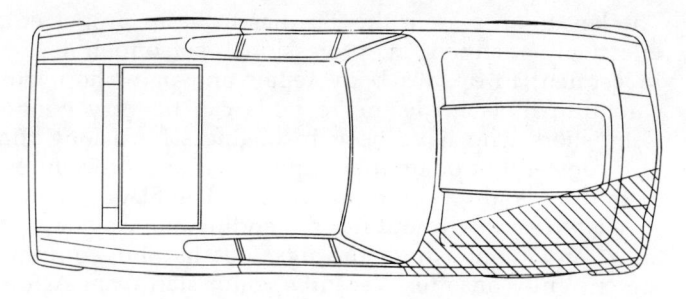

Fig. 22. 40% offset damage

Size of Vehicle	Small (Fiesta)	Medium (Escort)	Large (Granada)
Panel			
Roof	0.4	0.5	0.75
F. Wing	0.2	0.25	0.3
F. Door	0.2	0.4	0.5
F. Panel	0.2	0.3	0.3
R. Door	–	0.4	0.5
Sill	0.1	0.1	0.1
R. Wing	0.2	0.2	0.3
R. Panel	0.2	0.2	0.2
Boot Lid	0.3	0.4	0.5
Bonnet	0.3	0.5	0.7
F. End	1.30	1.8	2.0
Side	1.30	1.8	2.0
R. End	1.30	1.8	2.0

Fig. 23. Guide to quantity of unthinned colour material used on vehicle panels (litres)

28.5 Thatcham times

Over the period of time that the BIA has been in operation a great deal of criticism has been levelled at the organisation by refinishers claiming that times established were 'unreal' and not representative. The author must declare an interest here, as he was involved with the BIA as it first opened and the major paint conglomerate for whom the author worked ensured that time was spent going through the repair jobs on the shop floor. Suffice to say that the 'Thatcham times' given in the handbooks produced were obtainable without undue or unrealistic working, provided that the refinisher was set up with the proper equipment and staffed by professional operators.

28.6 Key to success

The key to success in the '80s has to lie in proper equipment effectively manned by professional personnel. It is a large investment to open a body repair and paintshop, but if this commitment is made the rewards can be very good indeed. Refinishers who have been in business for a long time must also constantly update their equipment and procedures as well as training and retraining personnel. The BIA at Thatcham will give advice and without recommending one way or another can show and demonstrate the jigs, booths and equipment that are currently on offer. Recently senior staff from Aston Martin Lagonda Ltd liaised with the BIA and as a result purchased two laser beam jigs for checking accident damaged structures and body conformity. This helpful cooperation is open to all and it is a fact that the BIA will do everything possible to help the refinishing industry become more professional and better equipped.

Poor workmanship cannot be blamed on poor tools in today's environment. Technology goes striding on and refinishers must keep up otherwise they will eventually fall by the wayside.

28.7 Training

The training of personnel can be organised by approaching any one of the major paint manufacturers who run courses of various lengths and at various levels for refinishers. Also, DeVilbiss Ltd at Bournemouth have an excellent training school where any refinisher can book a course for his staff. There is a large range of courses and training to be had, which serve to make the refinisher and repairer more professional, which is the aim of the BIA.

29
Thinners

29.1 Use:
To reduce primer or colour to correct viscosity for spraying. Two-fold operation:
A. Paint vehicle
B. Allow colour to flow – pigment dispersion.

29.2 Correct thinners:
Incorporates the optimum solvent level in formulation – delicate balance – low boiling solvent causes rapid dry and orange peel – high boiling solvent remains in the film to give flow and gloss potential.

29.3 Cheap thinners:
Usually reclaimed or unbalanced – use as gun wash.

29.1 Use

Solvents are used in the manufacture of paint to break down the pigment and binder into a more fluid material for tinting, controlling and finally canning.

A great number of thinners are available to the refinisher for thinning the paint to the correct viscosity at which it can be properly applied. The main differences lie in the evaporation rates, the ability to dissolve the paint vehicle, and thin the paint.

29.2 Correct thinners

Every thinner recommended for a particular type of paint should be a blend of solvents carefully formulated to have the correct dissolving power for the paint vehicle, and to have an overall evaporation rate which will allow the quickest drying time consistent with good flow.

The solvents already used in the manufacture of the paint must be taken into account when the thinner used to reduce it to application viscosity is formulated.

Set out below is the formulation for a cellulose thinner and it can be seen clearly that the blend is well balanced to achieve the breakdown of the cellulose colour.

Formulation of cellulose thinner:

	%
MIBK (methyl isobutyl ketone)	10
Isopropanol	15
Cellosolve	15
MEK (methyl ethyl ketone)	30
Xylene	20
Toluene	10
	100

29.3 Cheap thinners

A 'cheap' thinner or 'unbalanced' thinner will give poor results if used instead of the manufacturers' recommended thinner.

30
Renovation of Second-Hand Vehicles

> **30.1 Renovation:**
> The number of second-hand cars on the market will grow as all the major manufacturers continue to over-produce. These second-hand vehicles should be presented to the would-be purchaser in the best possible condition.
>
> **30.2 Operation:**
> To renovate the vehicle it is necessary to follow a full sequence of operations to bring the vehicle up to a first-class condition.

30.1 Renovation

The number of second-hand cars coming onto the market is increasing at the rate that new cars are being produced by the European motor manufacturers. The market is over-loaded and it is likely to continue to be so for some considerable time to come. Before the resale of a second-hand car, some attention must be given to the vehicle to bring it up to a saleable condition on the forecourt. Properly serviced, vehicles of today are unlikely to need much mechanical attention, but the body and interior will require examination and attention to the paintwork. The object of care renovation is to clean and refinish the vehicle as necessary to a high standard for close examination at the point of sale.

30.2 Operation

A. Wash and clean the whole vehicle thoroughly.
B. Lightly compound the exterior finish to clean off road film.
C. Clean all door returns and shuts with T Cut or similar cutting polish.
D. Examine the car panel by panel for surface damage, cuts and scratches as well as stone damage.
E. Prepare a silhouette for costing and work to be carried out.
F. Prepare areas of damage or blemishes for painting.
G. If the vehicle requires more than a third of total panel area for painting consider a total respray. This can be quicker and less expensive in the long run. For example, there may be customer

30.2 Vehicle Painter's Notes

      resistance to a partially-painted vehicle with metallic finish where fresh paintwork can be detected.

- H. Paint the vehicle as necessary.
- I. When fully cured out, compound the whole car lightly to take off the 'newness' look of fresh paint.
- J. Clean off all overspray, paying attention to chrome or other body fittings. Black out any overspray that has reached the wheel arches.
- K. If the road wheels are to be sprayed silver ensure that the correct wheel silver is used. Mixing enamel silvers are not suitable.
- L. Clean the tyres and use the correct blacking material if it has been decided to black the tyres.
- M. Clean all the glass both inside and out with a first-class cleaner. Clean windows help sell a motor vehicle.
- N. Clean the interior with recommended fabric or leather cleaners and replace rubber mats if necessary.
- O. Be critical of the final job and ask yourself if you would buy the car at the asking price.

31
Vehicle Care in the Paintshop

31.1 Vehicle care:
When a damaged vehicle is brought in it must receive proper care and attention. It is advisable to sheet up vehicles if they are to be left for any length of time.

31.2 Interior:
It is important to use floor mats and seat covers.

31.3 Exterior:
Care must be taken that no harmful substances contaminate the paintwork, such as anti-freeze or battery acid.

31.1 Vehicle care

When a damaged vehicle arrives in the paintshop for repair work it is often the case that the bodyshop have already carried out repairs. Unfortunately, when vehicles are stripped out they seem to attract a lot of dust and dirt from some bodyshops who are not quite aware of the problems that a dirty or damaged interior can cause.

It is normally left to the paintshop to finally valet the car after it has been completely repaired.

It is important to work on a clean car, and if it is clean when it arrives, to keep it in that condition. Various steps should be taken to ensure this.

31.2 Interior

    A. Use seat covers.
    B. Use floor mats.
    C. Clean the interior.

31.3 Exterior

On the exterior of the vehicle ensure that after painting no spillages take place. Remember that:

    A. Brake fluid removes paint.
    B. Anti-freeze stains and removes paint.

31.3 Vehicle Painter's Notes

C. Petrol and oil soften paint films and discolour them.
D. Battery acid corrodes.

It is important that the paint foreman makes all the other trades aware of the damage that can be caused to a customer's car whilst in the care of the refinisher or main agent.

It is also important that management support the paint foreman in ensuring that all staff are careful with vehicles.

32
Touch up

> **32.1 Introduction:**
> Sometimes necessary to touch up chips in paintwork.
> Door edges – stone chips on bonnet or boot – wing edges – wheel arches.
> Very small areas do not warrant spraying.
>
> **32.2 Method:**
> Rub out where possible using dry paper – touch in primer and filler – allow to dry – touch in with unthinned colour – build up coats – allow to through dry compound and polish.

32.1 Introduction

Sometimes it is necessary on a job to touch up very small areas, edges or stone chips. It may be considered that applying colour with a spray gun is unnecessary. It is quite permissible to touch up using a small fine brush and then to flat and polish out the blemish.

32.2 Method

Flat out where possible using wet and dry paper in the dry form. Touch in the area with primer filler that has about 20% thinners added to it. This will ensure that it dries out quickly and gives a massive spot build.

Allow to dry and then follow on with colour also thinned to about 20% thinners to 80% colour. (This applies to nitrocellulose and acrylic lacquers.) Touch in the colour and allow a flash off time between applications. Allow the material to through dry and then lightly wet flat with P 1200 wet and dry paper. Then compound and polish the area.

This can work very successfully and will improve an edge or chip so that the blemish is totally lost.

33
Lining

33.1 Types:
Stick on – brush on – spray on.

33.2 Stick on:
Fablon – sticky backed – peel off backing and pressure fit – remove with heat lamp – oven run.

33.3 Brush on:
Tape lined on pressure fit – centre section drawn out – cellulose or synthetic brushed in.

33.4 Spray on:
Tape lined on pressure – mark up and spray section drawn out – metallic finish.

33.5 Fablon transfers:
Ford JPS Escort 2000 – Daf Marathon – Renault R5 – take time and care.
JPS Capri line up wing edges – wheel arch tapes – handed 4 different.

33.1 Types

There are three types of lining that are currently used on vehicles for decoration and they are:

A. Stick on fablon sticky-backed.
B. Brush on.
C. Spray on.

On top of that there are now a whole range of transfers for fitting to the vehicle such as the Ford Capri JPS, the Ford Sierra XR4, the Renault R5, etc., and these devices attract customers to the vehicles and hence increase sales. However, after accident damage these have to be replaced after painting.

33.2 Stick on

A. Ensure that the car is clean and the paint surface is up to shop temperature (68 to 72° F).

B. Peel off part of the backing and set a point at the front or rear of the vehicle if a whole side is to be lined.
C. Stick the line to the point and place masking tape over that to hold it firmly in place.
D. Unroll the tape and cut it at the desired length.
E. Pull off the backing holding the tape away from the vehicle.
F. Either from the rear or front of the vehicle and holding the tape away from the job still, sight down the length of the vehicle and slowly bring the tape to the panels. This will ensure a straight line.
G. Rub along the line with mutton cloth and put pressure to the panel.
H. Remove the outer clear covering of the tape pulling it back along itself so as not to pull it off from the panel surface.
I. Cut the tape line in the door and wing apertures and fold it round behind the panel.

To remove this stick-on line easily apply heat from an infrared lamp or place the car in a low bake oven. The line will peel off quite easily after the application of heat. Do not attempt to paint over it and then stick another line on top.

33.3 Brush on

A. Ensure the car is clean and the surface free from any contamination.
B. Apply the tape in the manner previously outlined.
C. Draw out the centre red section of the tape.
D. Brush in a QD synthetic colour with a fine brush.
E. Allow overnight to dry out.
F. Remove the two outer edges of tape carefully.
G. Do not attempt to polish or interfere with the painted line.

33.4 Spray on

A. Ensure the car is clean and free of contamination.
B. Apply the tape in the described manner.
C. Pull out the centre section.
D. Mask up at least 12 inches either side of the line.
E. Spray in the metallic coach line.
F. When through dry remove the masking and line tape.
G. Polish if necessary to ease an edge.

33.5 Fablon transfers

These are very expensive items and must be treated with great care. Normally they have an original manufacturers' part number and this must be identified to obtain the correct part. When

33.5 Vehicle Painter's Notes

attempting to put these transfers on it is advisable to have help, both for checking and for fitting.

Many of the parts are handed and are different front to back, such as wheel arches and squared finishing sections on the Ford Capri JPS. If an error is made it can be costly to rectify.

A. Ensure the panels are clean and free of contamination.
B. Ensure the panels are up to shop temperature.
C. If no guide line is present, i.e. the next panel with original lines, then set up the transfer without removing the backing against the job and attach it with a corner of masking tape. Lay up the whole side in this manner to ensure a good fit and that all the correct parts are to hand.
D. Starting with the front wing remove the backing and place into position and smooth down with mutton cloth.
E. Using the front wing transfer as a datum line work back along the car.
F. When all the lines are in place remove the top clear cover from the line and press down hard with the mutton cloth.
G. Allow the vehicle to stay in warm conditions overnight to allow the adhesive to take up into the paint film.

Do not hurry this job. It takes a great deal of care and patience to obtain a first-class job. Extra time should be allowed for the paintshop to carry out this operation.

34
Industrial Fall-out and Other Contamination

> **34.1 Contamination:**
> There are three types of contamination sources.
>
> **34.2 Industrial fall-out:**
> Varying forms of this contamination, including sulphur fall-out from factory chimneys, iron and steel particles borne on the wind and chemicals discharged by industry.
>
> **34.3 Acid rain:**
> Acidic and alkaline fall-out causing contamination. Base pigments in the paint film come under attack from this pollution.
>
> **34.4 Agricultural sprays:**
> Certain insecticides that are carried from fields onto roads and motorways. Chemical attack discolours the film and aids breakdown.
>
> **34.5 Prevention:**
> Regular washing and general car care help to keep contamination at bay.

Modern car finishes are very resistant to all normal forms of atmospheric attack. Provided a simple maintenance procedure is followed they will retain their gloss colour and protective properties throughout the life of the vehicle.

However, car finishes are not chemically resistant. Severe local contamination of an acid or alkaline character can occur. If it is left in contact with the paint film for any length of time it may cause pitting and colour change.

34.1 Contamination

Contamination may come from three sources:

A. Airborne industrial fall-out and acid rain.
B. The vehicle itself.
C. The highway and adjacent ground.

34.2 Industrial fall-out

The term industrial fall-out was first used to describe minute particles discharged from the chimneys and workshops of the iron and steel producing and fabricating industries. Fall-out may be heavy near manufacturing industries but it is in no way confined to such areas. Strong winds can carry the effluent many miles. Railway networks are another source, the braking of rolling stock producing a discharge of iron particles.

In the presence of moisture and particularly if they are in the least bit magnetic, the particles will become attached to the paint surface. Soon they may discolour (rust coloured spots) and pit the surface of the paint. If allowed to remain they may penetrate right down to the primer or even the metal substrate itself. Regular and frequent washing is the best safeguard against attack, but if attack does occur:

A. Light contamination, when the particles are not embedded in the film, may be removed by compounding and polishing.
B. Heavy contamination can be removed by a chemical wash using a 10 per cent (54 grams) solution of oxalic acid to $\frac{1}{2}$ litre of water. Apply the solution taking care to prevent the solution from running behind mouldings, etc. It is important to keep the surface wet and active by several applications of the solution over a period of 15 to 20 minutes. Thoroughly wash off all traces of the solution and dry off. Then polish as necessary.

34.3 Acid rain

Acid rain is the term given to rain containing effluents from manufacturing chemical industries, particularly power stations. Some of the effluents may be acidic or alkaline in the presence of water, e.g. sulphur dioxide will dissolve in water to give an acidic solution, whilst a mixture of cement dust and water is strongly alkaline. Such effluents will attack paint films. The attack may take the form of discoloured spots due to attack on the pigment (e.g. some reds will develop a blue tone if attacked by acids and a brown discoloration if attacked by alkali) or distortion of the paint vehicle itself. Brunswick greens are well known for their proneness to attack by alkali (yellow discoloration).

Some pigments are more prone to attack than others. The aluminium flake in metallic paints is particularly prone to attack by both acid and alkali. Basecoat and clear metallic finish has the advantage that the clear coat shields the aluminium from the contaminant. But even clear paints may be attacked, losing transparency and/or gloss. Air dry paints, particularly when new, are more vulnerable than stoved finishes, but become more resistant to attack as they age. Of the present range of refinish topcoats, best resistance to acid rain attack is shown by the polyurethane paints.

34.4 Agricultural sprays

Contamination also comes from agricultural and horticultural sprays. Certain insecticides such as DDT, dieldrin and Malathion can cause spotting and pitting or blistering of paint in the presence of moisture and warm sunshine. Some concentrated herbicides are literally paint removers. Bird droppings can cause severe distortion and/or discoloration of paint films.

In the United States, the residues of some types of dead fly, when baked on by the sun, have caused discoloration and cracking.

Contaminants from the car include petrol, oil, grease, brake and de-icing fluids, which may result in staining and/or breakdown of the film. All these contaminants may be present on the highway itself, as may exhaust wastes, tar and other road building materials and residues from various spraying operations, e.g. creosoting of fences, crop spraying, etc.

Tar, bitumen and grease stains can be removed by wiping with white spirit or similar suitable solvent, but light contamination from acid rain needs compounding and polishing. Heavier contamination may require that the vehicle be painted – wash thoroughly clean with a solvent based water miscible cleaner, wet flat, dry off, spirit wipe and repaint. In extreme cases it may be necessary to strip to metal and repaint.

34.5 Prevention

Regular and frequent washing is important in preventing chemical attack, particularly when the car is new or has been freshly painted. Some degree of protection may be obtained by regular cleaning and waxing. Better protection will be provided by polishes containing silicone, but there is no guarantee of freedom of failure by way of contamination.

Set out here is a table of some of the sources of contamination and the effect they have on a paint film, and the colour.

34.5 Vehicle Painter's Notes

SOURCES OF CONTAMINATION		EFFECT OF CHEMICALS ON CERTAIN COLOURS		
Industry	Chemicals used	Colour	Appearance	Chemical cause
Fertiliser	sulphuric acid nitric acid phosphoric acid superphosphoric acid nitrogen compounds	Yellow	white spot	hydrochloric acid – muratic acid
			dark brown spot	nitric acid
			red spot with film degradation	sodium hydroxide – caustic
Cement	potash–potassium hydroxide lime–calcium hydroxide	Non-metallic, medium depth blue	slight whitening	nitric acid
Paper mill	sulphuric acid lime		slight whitening with film degradation	sodium hydroxide – caustic
Distillation of hardwood	acetic acid		spotty blistering	acetic acid
Textile fibres	caustic soda other caustic compounds nitrogen compounds	White	pink	nitric acid
			pink colouration with film degradation	sodium hydroxide – caustic
Petroleum	sulphuric acid		spotty blistering	acetic acid
Chemical	almost any possibilities including solvents		yellowing (acrylic lacquer showed no discolouration)	ammonium hydroxide
Dyes	sulphuric acid hydrochloric acid nitrogen compounds hydroxides nitric acids	Medium depth blue	slight light blue spot	hydrochloric acid
			dark blue spot	nitric acid
			deep purple spot with film degradation	sodium hydroxide – caustic
Pigments, paints varnishes	sulphuric acids			

Industrial Fall-out and Other Contamination 34.5

SOURCES OF CONTAMINATION		EFFECT OF CHEMICALS ON CERTAIN COLOURS		
Industry	Chemicals used	Colour	Appearance	Chemical cause
Lacquer printing inks	caustic-sodium hydroxide		spotty blistering	acetic acid
Soap and detergents	caustic soda potassium hydroxide caustic-sodium hydroxide			
Steel, copper, lead, zinc	caustic-sodium hydroxide			
Tin and mercury	sulphuric acid			

35
Spraying Shapes

> **35.1 Object:**
> To coat evenly any shape to a uniform depth with a continuous film of material.
>
> **35.2 Gun position:**
> Gun must follow contours of panel at an even distance throughout.
>
> **35.3 Spraying corners:**
> Reduce pressure spraying into corners – avoid heavy build-up – avoid blow-back of material.
>
> **35.4 Versatile:**
> Alter gun position – pressure – set up – viscosity to suit condition.

35.1 Object

When using a spraygun to paint a vehicle the basic object of the exercise is to coat up evenly any shape to a uniform depth with a continuous film of material.

The only way that this can be successfully achieved is to ensure that the gun traverse remains constant in relation to the panel that is being sprayed, i.e. the gun remains at a constant distance from the workpiece, between 6 inches and 8 inches, and at right angles to the surface of the panel. In the case of primers, fillers and straight colours an overlap of 50% is usual. In the case of metallic finishes this can be as high as 80%.

35.2 Gun position

The gun must follow the contours as already described, with little or no deviation from that. Constant practice will ensure that these tolerances can be maintained.

35.3 Spraying corners

When spraying into corners, particularly door shuts or returns, it is advisable to reduce the pressure at the gun considerably to lessen the blow-back. However, this tends to lead to poor atomisation and to overcome this the paint should be further thinned. However, it is possible with skill to paint the shut by dusting down the panel and allowing the paint to strike at a more oblique angle. For door shuts and corners that are not in the prime site for both viewing and weatherability, this is acceptable.

It is important to minimise the build-up of paint on either side of an enclosed section otherwise runs and sags will be produced. A lighter application is sensible. Return to the area and build up material slowly.

35.4 Versatile

Always be ready to alter gun position when spraying and develop a more versatile approach to the job. As long as the end result is an even film build to a required depth then the correct result has been achieved.

Do not hesitate to vary pressure, gun set up and viscosity to suit a certain situation. Generally this must be for panels and interiors that cannot be seen as such and are protected from the weathering process that the vehicle suffers on the cosmetic panels.

Bearing in mind the correct procedures and carefully following the paint manufacturers' recommendations, a painter should develop style and versatility so that he may improve his skills, aptitude and approach to the job.

36
Fault Finding

36.1 Visual observation:
Careful observation of paint film – use magnifying glass – do not guess – ask questions – use depth gauge.

36.2 Examination:
Decision to disturb paint film taken – proceed with care – all damage must be refinished – flat down to bare metal and examine with magnifying glass.

36.3 Decision:
Decide on remedy or rectification after careful thought. Call in the paint manufacturer for help.

So often when a vehicle is brought into the paintshop for fault analysis, when everyone from the general manager down suddenly becomes a paint expert, basic rules of observation and careful thought appear to be forgotten.

Paint finish on motor vehicles is a complicated subject and only now are management and users beginning to realise this. In many cases there is no straighforward or obvious reason why a fault has occurred and only careful examination and a further process of elimination might clarify the position. It is often said that people only see what they either want to see or expect to see. It is important to take all the emotion out of the situation and try to clinically observe the fault and come to a considered opinion. This cannot be done rationally with a customer at one's elbow insisting that he cleans the car every week come rain or shine, and management anxious to please accepting liability for a fault that may have been induced by the owner. When examining a fault the following procedure should be followed.

36.1 Visual observation

Gather all the information possible about the vehicle.

A. How old is it?
B. Is it original finish?
C. Where is it garaged?
D. Is it parked outside all the time?
E. Does the owner work in a blast furnace or steel mill?
F. Does he park it under trees?

G. Talk to the owner and find out what he cleans the car with and the frequency of the cleaning. Does his wife or son clean the car? Do they use an abrasive cleaner? Does he use a car wash regularly? Do they cover the car with a plastic cover so that humidity becomes a problem?
H. Does the owner live by the sea?
I. Does he normally care for his car?

The list of questions is almost endless but from this some sensible understanding will evolve. Ask questions and *keep on* asking.

36.2 Examination

When examining the paint film and the fault ensure that the vehicle is clean and there is no dirt or other contamination to lead the examiner astray.

Observe the fault in daylight and with the aid of a microscope, a small pocket variety giving $50\times$ or $60\times$ is invaluable, or a good magnifying glass.

Look for disturbance in the film and particularly contamination from industrial fall-out which is becoming ever more of a problem to the finish on a motor vehicle.

Establish whether the fault has penetrated the paint film from without or whether it is erupting from below the top colour coat.

36.3 Decision

Carefully weigh up all the possibilities of how it has occurred and why it has occurred and then by careful elimination start discounting those reasons that may not fit the picture. Sometimes it is necessary to leave the vehicle and contemplate overnight, and when an impasse seems possible do not hesitate to call in the technical representative from the paint manufacturer. They are extremely helpful and provided you have used their materials exclusively on the job, or the car is new to you, they will still inspect it and give you their opinion. Behind them they do have the extensive facilities of a laboratory.

When examining the film the decision to disturb the film may be taken. Do so remembering that it has to be put right later.

37
Paint Thickness

37.1 Depth:
All paint films should be uniform depth – correct surface coating ensures maximum durability.

37.3 Measurement:
Readings can be obtained by use of a depth gauge. A magnetised pull-off instrument – measured in thousandths of an inch or microns.

37.3 Paint thickness:
High stoved original finish – approximately 4 to 5 thou – over 12 thou gives recoat problems.

From early days in the automotive industry it has been the norm and accepted by the public at large that many coats of paint applied laboriously one upon another was the only way to obtain a splendid finish and an almost infinite life of that paint finish. When vehicles were built on chassis with the bodywork assembled on the chassis later it was possible to paint a car in this fashion. Remember, however, that coat after coat was applied and then flatted well back in an attempt to get a surface. Most of what was applied was later removed. The paint technology in those early days was crude in comparison to the products on offer today.

37.1 Depth

Once car manufacturers went on to monocoque construction new developments were called for. If a modern motor vehicle was to be painted up with massive film builds even with modern materials, a breakdown in the film would occur.

37.2 Measurement

With Ford, GM and BL working to top limits of up to 8 thou (200 microns), to go much above that will begin to seriously stress the modern finish. Above 12 thou (300 microns), the chances of paint fracture on a monocoque body are likely. A modern motor vehicle flexes and moves continuously as the car is driven. The paint finish will not move in the same way if it is too thick.

37.3 Paint thickness

Before refinishing a vehicle, where there is some doubt as to the film thickness, it is well worth taking a reading using a Tinsley depth gauge. This is a magnetic 'pull-off' type which will give the reading both in thou and microns. If there is doubt do not hesitate to strip a panel back to bare metal. It is safer and quicker in the long run.

38
Infra-red Heat Lamps

38.1 Background:
Infra-red curing has played an ever-increasing role in original finish and refinish vehicle painting. All major motor manufacturers are using this form of curing at various stages.

38.2 Reasons for use:
A list of reasons for the use of this equipment, for example, speed of operation, and cleanliness.

38.3 Use on complex shapes:
The radiant energy allows complex shapes to be fully cured out

38.4 Selection of equipment
Wavelength selection is necessary for the task to be attempted.

38.5 Factors in curing:
A list of factors to be taken into account for correct and effective curing of a paint film.

38.6 Types:
The set-up of banks of emitters to complete the job.

38.7 Use:
The use of infra-red on all types of finish, from air dry to low bake.

38.1 Background

Infra-red heat lamps have played an ever-increasing role in the finish and refinish of the motor vehicle. All the major manufacturers are now using infra-red curing in pre-colour areas, final finish or rectification line procedures.

The refinishing industry has developed the use of infra-red for part repairs and localised finishing, and it is rare to find it used in the full respray low bake conditions.

Most vehicle and industrial finishes dry by a process of solvent removal or by polymerisation, both of which can be accelerated by the application of heat. Motor vehicle finishes having desirable properties such as hardness, flexibility, adhesion, chemical resistance and freedom from blemishes are easily obtained using electric infra-red ovens.

Stoving temperatures are rapidly attained and the results compare very favourably with those achieved by alternative heating methods such as convection ovens.

38.2 Reasons for use

The wide variety of electric infra-red equipment now available is being exploited increasingly in the industrial finishing field for the following reasons:

A. Rapid heat transfer to panels shortens treatment times and increases throughput.
B. Rapid heat up and cool down of heaters saves idling losses and therefore protects finished products during line stoppage.
C. Highly suitable for mass production operations.
D. Simple and accurate temperature control.
E. No contamination from products of combustion.
F. Equipment is relatively inexpensive and easy to install.
G. Energy is used efficiently.
H. Simple maintenance.
I. Compact equipment saves factory space.
J. Cleanliness during use, which is particularly important for a refinishing shop.

38.3 Use on complex shapes

The radiant energy from an infra-red source, the emitter, is directed towards the product to be heated which ideally should have a large flat area and small thickness. However the emitted energy can heat the whole surface of a three-dimensional product if it is suitably jigged.

In the metals finishing field, conduction of heat throughout a complex product shape can reduce temperature gradients and play an important part in ensuring that shadowed corners or interior coated surfaces are properly heated.

Convected heat can make a secondary but important contribution to the heating process, depending upon the type of emitter used. The energy emitted from the infra-red source increases rapidly with its absolute temperature, obeying, in theory, the fourth power law. The wavelength of the radiant energy depends upon the temperature of the source, which for industrial purposes will be in the range of 600 to 2200° C.

38.4 Selection of equipment

The selection of infra-red equipment depends largely on the physical parameters of the process, but particularly on the nature of the material to be heated. Its colour, roughness, and chemical structure, collectively referred to as its 'absorptivity' or 'emissivity', have a marked effect on the wavelengths absorbed.

It is therefore desirable to match the infra-red heating system with the coating formulation to ensure the best results in terms of energy economy, productivity and surface quality.

Thermosetting paints require a minimum temperature to cure in a commercially convenient time and this can be generated by infra-red radiation.

IR emitters generate radiant energy in the 0.7–4 μm region and are characterised by their peak emission, short, medium and longwave. Medium and longwave are not as 'peaky' as short. Organic and inorganic paint media have specific absorptions in the near IR due to overtones of principal absorption, molecular vibrations and charge transfer phenomena: these absorptions are dissipated as heat.

Cure of paint media may be isothermal (nominally) or by superheat raising the temperature until cure occurs. The great advantage of IR heating is the fast temperature rise, due to the large temperature difference between emitter and object. Superheat is a more effective way of using IR.

38.5 Factors in curing

Factors affecting IR cure include:

A. Emitter type.
B. Emitter to product spacing.
C. Substrate thickness.
D. Substrate texture.
E. Paint colour.
F. Total film thickness.

Dark colours heat up faster than light colours. The rate of heating is related to the red absorbtance of colours, e.g., white may attain 72° C in the time a dark blue takes to attain 87° C.

A thicker white film heats up slightly faster than a thinner one, but this difference will not occur with a black film. A rougher texture to the metal substrate may also cause a white film to heat up more quickly, again a black film is not affected.

In isothermal cure, a saving in time to cure is confined to the heating up time. In superheat cure, considerable savings in oven dwell time can be made, for example an alkyd MF paint stoving at 120° C for 30 minutes in the isothermal mode can be cured in about five minutes in the superheat mode, depending on the kind of emitter and other factors.

Infra-red cure is a valuable means of speeding up paint stoving, it is clean and the ovens and arches are very compact. It is very important to ensure that proper advice is taken before the installation of a heater, arch or even an oven, to ensure that the correct radiation is used to obtain the best results.

38.6 Types

Systems include lamps or emitters in banks of two or four or multiples. Complete contoured arches can be installed and these can be used for the complete cure of the whole vehicle or part of the vehicle after colour application. They may also be used as part of the preparation process to ensure that moisture is driven from the substrate prior to colour application in a suitable spray booth. A medium wavelength of 3.0 µm is particularly suited to drive out moisture from a semi-permeable membrane and so is recognised as a major step in the eradication of faults caused by moisture, such as blistering and micro-blistering.

38.7 Use

On all air dry or low bake materials. Small lamp units are ideal for localised repairs and they can be used anywhere in the paint shop. Ensure for safety that no spraying is carried out with the lamp or emitter on, and that no water is splashed on the unit as it will shatter the bulbs. Check the distance of lamp to workpiece as the temperature rises very rapidly.

39
Spot Repairs

> **39.1 Object:**
> Often it is necessary to carry out a small localised spot repair. It is a difficult area of refinishing where experience and knowledge are necessary.
>
> **39.2 Procedure:**
> A full procedure list of how to refinish a spot repair in the paint film.

39.1 Object:

Often it is necessary to carry out a very small localised repair, and this is usually either due to slight damage that has occurred or for warranty purposes, e.g. dirt inclusion in a metallic finish. It is important to keep in mind that a small localised repair, although called for, may not be the best way forward in attempting an invisible repair. This is a difficult area where the operator must make the decision on how best to proceed. A great deal depends on the position of the fault, for example, in the middle of a bonnet, a possible complete refinish of the total bonnet may be the best course.

39.2 Procedure:

The procedure for a successful refinish is as follows:

A. Check the complete panel for any other damage or scuffing.
B. Decide whether to attempt a spot repair or completely finish the panel.
C. Clean off and lightly compound the whole panel to remove road grime.
D. Make a good repair. If the body shop have been involved, check the work standard very carefully. Remember the customer knows exactly where the damaged area is and it will come under very close scrutiny.
E. Wet flat and feather out to maintain good control of the original paint surface. DA sander preparation is not advisable unless a deep rust spot is the cause of the repair.
F. Spot prime and then apply filler coats.
G. Guide coat and wet flat with P 600 and use a block.

Spot Repairs 39.2

H. Be accurate and ensure that the blend in of filler to original finish is very smooth.
I. Check the colour match very carefully before applying the colour.
J. Spray the area overlapping each coat to soften the overspray from the previous application.
K. At this point either: (a) Spray overthinned colour onto the final edge of the repair to soften it. Then after a complete and full dry through compound and polish the repair; or (b) wait until a full and proper through dry has occurred and then wet flat the whole repair area with P 1200 and then polish with compound and finish off with T Cut.

There is a different process for spot repairing metallic finishes and that is dealt with separately.

The most important thing to remember with straight colour repairs is to ensure not only a good body repair if it has been necessary but that the colour match is absolutely spot on. If the customer can see it he is likely to ask for the repair to be done again. Again, care and cleanliness at each stage of the process will ensure that a first-class result is achieved.

40
Mixing Schemes

40.1 Types:
There are two types of mixing scheme generally in use. The volumetric and the gravimetric.

40.2 Operation:
Operating a mixing scheme saves both time and money, as well as giving the latest colour to the user at all times.

40.3 Updating:
As new colours are devised, the user is updated by the paint manufacturer.

40.4 Variants:
Colours that may vary on the motor manufacturers' production line can be offered as variants to the user with new formulations.

40.5 Housekeeping:
Good housekeeping ensures that the equipment and system remain 100% operational.

40.6 Mixing procedure:
The full listed procedure for dispensing colour.

Because of the number of colours being brought onto the market by the car manufacturers the need for some form of mixing scheme was soon evident.

40.1 Types

There are two main types of mixing system in use, and they are:

- **A.** The volumetric – volume of mixing colour.
- **B.** The gravimetric – weight of mixing colour.

The most popular type of mixing scheme is the gravimetric and it is widely used throughout the UK and Europe.

The weight mixing scheme has been fully developed to cope with the avalanche of new colours that were introduced by the motor manufacturers and made worse by the number of Japanese and European imports. The use of the mixing scheme is one of the most significant developments in the vehicle refinishing industry. To dispense colour by weight brought forward many problems.

Firstly, the accuracy of the equipment itself. It had to have the ability to accurately weigh down to 0.5 gram over a weight range of 0.5 to 5,000 grams, which is approximately the weight of 5 litres of colour.

Secondly, the staining strength of the mixing enamels had to be constant, for repeatability, and not to suffer with batch wander on any area of standard.

With volume mixing the enamels there is a tendency for less accuracy as the operator is required to finally 'tint' the colour by eye as necessary.

Mixing by weight is at least five times more accurate than some volumetric mixing schemes. Computers now have made the matching of new colours a matter of routine, and the coding of the weight formulation a straightforward affair.

Paint manufacturers offer a range of products in weight mixing form. Nitrocellulose, which is still widely used in the UK, acrylic finish which possibly is not so popular, the new two-pack isocyanate finishes which are used extensively in Europe and are gaining rapid acceptance in the UK market now, and finally synthetic finishes.

40.2 Operation

Operating a weight mixing scheme saves time and money. The time saving revolves round the ability to match any colour, in any quantity, there at the point of use without having to refer outside the paintshop.

The money saved is in the stockholding. Before the mixing schemes, colours were bought and delivered daily, from a supplier. Usually, the paintshop over-ordered in case of faults or rework. This material went into stock, 10 tins became 20, 20 became 40 and so on. The stock holding of colour became very expensive in most paintshops. Most of it ended up being thrown away. With the mixing enamels necessary to operate the scheme only approximately 40 need to be held in stock, although this does vary a little from paint manufacturer to paint manufacturer.

40.3 Updating

As stylists and designers develop more colours, the paint manufacturers can offer the refinish trade the colours instantly by updating the microfiche formulas. Weight mixing is here to stay, and the back-up that these systems can give the painter is immeasurable.

40.4 Variants

For example, many colours vary on the production line due to colour batch wander, stoving times or different suppliers. As many as seven variants can occur for one colour and the weight mixing scheme offers formulations for each variant.

The ability to match these colour variants means more accurate refinishing and hence faster turn-round. The fact that the formulation is known to the painter also means that, should the decision be made to alter a colour, the knowledge of precisely what enamels are in the colour make up allows him to tint with those enamels. The more mixing enamels that are present in a colour the 'dirtier' it becomes. It is therefore important to ensure that no new enamel is introduced to the formula when tinting.

40.5 Housekeeping

It is essential to maintain good housekeeping in the paint mixing bay and here are some basic guidelines:

A. Keep up to date with all the new microfiche and relevant colour information.
B. The mixing scheme can only be expected to operate accurately provided that:
   - The basic enamels are well stirred by hand or by machine mixing head.
   - The scales are kept clean and regularly serviced by the makers or appointed agents.
   - The microfiche is kept up to date.
   - All the equipment, including benches, is kept clean.
C. The mixing room or bay should not suffer extremes of temperature as this can affect the performance of the enamel in the can, i.e. variation of viscosity levels.

40.6 Mixing procedure

When more than one topcoat type is available, it is important to select the correct microfiche formula. Check that this is the latest fiche available.

Before mixing, the stirring head should be run for the correct period both first thing in the morning and then in the afternoon (approximately fifteen minutes each time).

Zeroing or tareing of the weight mixing or volume mixing equipment should be carried out carefully.

All mixing formulae shown on a microfiche indicate accumulated values, eliminating the need for any calculations while mixing. This allows careful concentration when pouring out the enamels into the tin. In the case of overshooting the required amount of paint with any but the first ingredient in the formula, do not attempt to 'adjust' the amount poured. A fresh start must be made to the whole procedure.

When the microfiche formula has been completed, remove the tin from the scales and stir it thoroughly by hand before thinning down with the viscosity cup for use in the spray gun.

Take care of the equipment and do ensure:

A. That all basics are thoroughly stirred before use.
B. That the pouring heads are kept clean from paint build-up.
C. That the scale pan is covered with masking paper.
D. That pourers are to the back of the can so as not to obscure the colour name and reference number.
E. A supply of clean rags and solvent.
F. That you do not use different basics in mixing.
G. That the weighing equipment is not affected by vibration or dust.
H. That empty mixing tins are stored upside down.

Used properly and cared for a weight mixing system will give years of trouble-free service.

41
Formulations

> **41.1 Object:**
> Due to numbers of colours on offer (13,000 plus) weight mixing by formulation is necessary. Formulations are key to making colour in amounts from ½ litre to 4 litre/5 litre.
>
> **41.2 Use:**
> Colour formulations used throughout the world for the refinishing trade. All products available to refinisher. Most formulations on microfiche system – updated monthly – all totals are cumulative – simple to use – accurate to ½ gram.

41.1 Object

To prepare colour formulations on microfiche enables the paint manufacturers to both update and review formulations easily and to then communicate new formulations to the end user. This service, that is provided by all major companies, is a most excellent demonstration of computer-controlled technology and how it can service a varied and fragmented industry. The formulations are set out in amounts from ½ litre to 4 or 5 litre capacity.

41.2 Use

They are accurate to within ½ gram and used properly with the weighing equipment will give total accuracy to the motor manufacturers' master panel. This does not necessarily mean a 100% match to the car in the paintshop, as many other factors come into play. The vehicle may have to have a colour variant formulation colour applied due to colour batch wander on the production line.

Great care must be exercised in the choice of colour before mixing takes place.

Microfiche formulations are normally updated monthly, with all products available to the refinisher.

All the totals are cumulative, so ensuring that an easy pouring operation can proceed.

Microfiches are simple to read and will show other background details such as motor manufacturers' code and date of introduction. Keep the microfiche formulae clean and dust-free, and ensure that as they become date-obsolete they are removed from the file and either destroyed or placed elsewhere. It is important to guard against confusion in a busy shop.

42
Microfiche

> Use of microfiche – updating – ease of operation.

Microfiche was developed for carrying large numbers of parts or formulations quickly recallable on a visual display. The system as it has been developed allows an instant update as new colours or paint types become available.

The paint manufacturers normally update each month in an attempt to keep up with the ever-growing number of car colours from throughout the world.

New paint finishes can be developed and then the colours made in that system are readily available to the user.

The microfiche is easy and simple to operate. Normally there is an index that will lead you to the colour formulation. Usually there is a numeric list and a motor manufacturer list set out alphabetically. After identifying the colour of the vehicle by use of a colour chip, it is simple to transpose the information on that chip to the microfiche. The formulation can then be copied out and taken to the mixing room where the colour can be mixed.

43
Basic Colours

> **COLOURS OF THE SPECTRUM: RED, ORANGE, YELLOW, GREEN, BLUE, INDIGO, VIOLET.**
>
> **43.1 Basic colours:**
> The development of tinters to strength-controlled basic mixing colours for the mixing systems gave the user the opportunity to mix over 13,500 colours from 40 to 50 base enamels.
>
> **43.2 Reduced tinters:**
> These are controlled strength materials with reduced staining strength, making it possible to get better colour dispersion and a more accurate tint.

43.1 Basic colours

To refinishers the term basic colours means the basic mixing enamels from which a huge range of complex colours and shades may be produced. The development of these basic colours and tinters has been the foundation of the highly accurate mixing systems now employed by many refinishers throughout the world.

Normally, something in the order of 40 basic enamels exist for each type of paint, e.g. nitrocellulose, two-pack isocyanate, etc., but this number does fluctuate from manufacturer to manufacturer. The number of so many car colours coming onto the market prompts further development and then yet another base mixing enamel. The secret of these controlled strength enamels is their staining strength when mixed together. In other words their effect colourwise on the total volume of paint.

43.2 Reduced tinters

It is necessary to get an even dispersion of pigment throughout the volume and therefore it is necessary to increase the volume but keep the staining strength the same, or the reverse of that. Therefore reduced strength tinters are available to ensure that a larger and therefore more measurable and more controlled amount of tinter is added to the paint mix to ensure an even colour change throughout.

43.2 Vehicle Painter's Notes

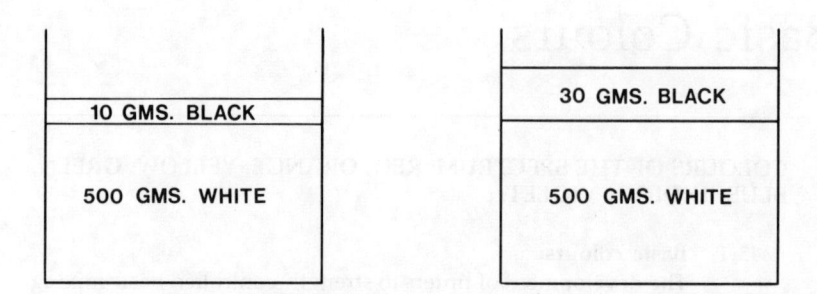

```
┌─────────────────────┐          ┌─────────────────────┐
│   10 GMS. BLACK     │          │   30 GMS. BLACK     │
├─────────────────────┤          ├─────────────────────┤
│                     │          │                     │
│   500 GMS. WHITE    │          │   500 GMS. WHITE    │
│                     │          │                     │
└─────────────────────┘          └─────────────────────┘
```

Fig. 24. Reduced tinter distribution compared with full strength mixing enamel

Figure 24 shows the positive and controlled effect of adding by weight and therefore by total volume a reduced strength mixing enamel and how in comparison with a full strength mixing enamel it gives a more accurate alteration through the entire volume of paint.

44
Colour Codes

44.1 Use:
Colour code systems used by all manufacturers of passenger vehicles, usually stamped on chassis plate – all paint manufacturers list codes against colour names.
Codes are accurate way of identifying colours and should be used in conjunction with colour chip.

44.2 Location:
Paint manufacturers give a location chart to pinpoint codes on every make of motor vehicle.

44.1 Use

All motor manufacturers use a code system to identify various parts of the car, and included in this is the code number for the body colour of the car. This code is normally stamped on the chassis plate and clearly identified. The paint manufacturers will list this number against their own in-house formulation number so that the refinisher can quickly cross-reference. The paint manufacturers have a large colour book that identifies vehicle manufacturers alphabetically and then colours of that vehicle manufacturer in the same way again.

As soon as a vehicle enters a paintshop the colour should be identified at once and this should be carefully checked. It may be that the colour is not one that can be mixed on a weight mixing scheme and therefore must be bought in from a supplier, or the mix formulation on the microfiche may state that the colour match is suitable for 'resprays only'. This may then figure greatly in the decisions that must be taken, such as, should the vehicle be resprayed? Should the complete side be painted?

When this occurs it makes sense to quickly make up the colour and spray out a 6 inch × 4 inch panel and check the colour against the car. Codes are a very accurate way of identifying colours and should be used in conjunction with the colour chip and the book of variants. Some colours have as many as seven variant colours and shades and careful work at this early stage can stop a wrong colour selection.

44.2 Vehicle Painter's Notes

44.2 Location

The location of the chassis plate is normally under the bonnet, but in some cases the colour code does not appear on this plate and is to be found in another place. The paint manufacturers give a diagram view of a vehicle with the colour code plate clearly indicated.

45
Basic Colour Match

> It is vital to ensure that the correct colour is selected and mixed before application to the vehicle. If the correct procedure is followed then it is possible to obtain a good match for every job.
>
> **45.1 Procedure:**
> Clean a panel adjacent to that which is to be sprayed for matching to. Check the paint manufacturers' colour chip to this panel. Spray correctly thinned mixed colour onto the test panel and check against the vehicle. Always allow colour to dry out and develop before deciding to tint. Use reduced tinters carefully.

Although the colour of a vehicle may be identified by the code on the chassis plate the colour may be at variance with that and it is essential to check the colour with a chip and the book of variants. There is a simple procedure to ensure that the correct colour is selected for the job.

45.1 Procedure

A. Clean the whole car before entering the paintshop.
B. Select two areas, a wing top and boot lid, and compound an area on these panels to remove dirt and road grime.
C. Observe the general condition of the vehicle and how well it is cared for.
D. If a wing or door is to be sprayed, compound the panels adjacent to it and observe if repairs have been carried out before on these panels.
E. Check the colour code and then the colour chip and variants to the vehicle. Carry out this visual check in daylight.
F. If in doubt, ask another colleague to look at the vehicle and colour chip, as often after a time of looking the naked eye becomes less sensitive to colour differences.
G. Make up the colour on the weight mixing scheme and then thin it correctly to the makers' recommendations.
H. Spray the colour out on a 6 inch × 4 inch primed panel.
I. Allow to dry and develop up. Remember the drying out as well as the very action of colour passing through a spraygun will cause various pigments to develop up in the film.
J. Check panel against the vehicle in daylight, and if necessary polish the panel with compound to take the 'newness' off the colour, prior to matching.

45.1 Vehicle Painter's Notes

K. If the colour is a perfect match then obviously the operator can proceed with the job.
L. If the colour match is not good then the need to tint must be established. If the colour is a metallic then other factors come into play. These are discussed at a later point.
M. Having established the need to tint the colour, return to the mixing scheme and observe the make-up of the colour.
N. Check the formulation carefully, and by a process of elimination decide what tinter is required to alter the colour. For example if the colour prepared is too light, and there is an amount of reduced black or reduced blue black, then it is in order to tint with these.
O. Before tinting a colour tip a small amount of colour into a tin for the purpose of tinting, so that any overshooting will not cause the loss of the total amount of colour. This can always be sprayed on as first colour coats and allowed to dry up.
P. Tint colour by small amounts and always spray them out on a 6 inch × 4 inch primed panel, and allow the colour to dry and develop.
Q. If the refinisher overshoots, then he can tip some of the main bulk of colour into the tinting tin to equalise out the fault.
R. Only as a very last resort move to tinters that are not in the original formulation. Any extra addition of tinter will certainly cause the colour to become dirty and then it is unlikely to match correctly.

There is great skill in colour matching by eye.

46
Static Electricity

> Static electricity can build up very rapidly in metal fixing containers, electrical equipment, bench equipment and solvent storage cans.
>
> Static discharge can ignite vapours and fumes causing explosion and fire. It is important to earth appliances, bench tops and use alloy stirring sticks.
>
> Keep tins sealed and remove all cloth and paper waste daily from the mixing room.

Static electricity is a hazard in the paintshop which is often overlooked. Static can build up very rapidly and cause an explosion or fire.

Metal mixing containers on a metal bench, coupled with electric mixing head equipment, all contribute to this serious hazard. The static discharge can ignite vapours and fumes causing an air flash and fire.

To minimize the risk from static discharge ensure that:

A. Electric appliances are earthed.
B. Bench tops are earthed.
C. Alloy stirring sticks are used.
D. Tins are always sealed and not left open.
E. Disposal of waste cloth and paper is completed daily.
F. Vehicles are earthed in the spray booth, especially when using two-pack materials.
G. Great care is exercised when using electrostatic spray guns, and that the equipment manufacturers' recommendations are carefully followed.

If the danger is properly recognized then a good refinisher can ensure that his paintshop is in no danger of this particular hazard.

47
Metallic Finishes

47.1 Background:
The increase in metallic colours has been remarkable and the trend is set to continue. Metallic colours enhance vehicles and are more attractive to the purchaser.

47.2 Manufacturing:
Manufacturing metallic colours is quite straightforward and paint manufacturers are able to make up any form of metallic colour.

47.3 Basic factors for matching:
It is vital to understand that the factors that govern a good metallic match are the colour, the appearance and the texture.

47.4 The colour:
This can only be changed by tinting.

47.5 The appearance:
This can be changed by gun technique.

47.6 The texture:
This can be changed by thinning ratios and gun technique.

47.7 Light reflection:
When light falls onto a metallic film it is reflected back by the aluminium flake acting like mirrors.

47.8 New developments:
New 'plastic' parts for cars will add to problems by introducing soft 'interface' materials.

47.1 Background

The increase in the popularity of metallic finishes has been quite remarkable. It has come about as the result of marketing trends to enhance the motor vehicle in terms of customer acceptance at the point of sale.

The major car manufacturers realised that it was the female who chose cars because of their colour. Having established that fact along with the persuasion that a wife could exert on a husband to choose a particular model because of its colour, it was clear that sales of motor vehicles and the finish and colour would progress hand in hand.

47.2 Manufacturing

To manufacture a metallic finish in simple terms is to add particles of aluminium flake to a basic colour. There is no adherent difficulty in metallising any colour.

The aluminium flake comes in various grade sizes depending on the type of ball mill that it has been ground in and the size of ball used in the grind.

The varying sizes of metallic particles are very important to the colour appearance and texture and must be accurately matched in any refinish material.

The full understanding of how a metallic colour is arrived at will help the refinisher to carry out an invisible repair to a damaged vehicle.

47.3 Basic factors for matching

So often a vehicle that has been repaired can be clearly seen by the owner to be different. Always he will claim the colour is wrong, but it may well be that other factors have caused the visible 'mismatch', and that the actual colour is correct. Originally metallic colours carried only one grade of metal flake in the colour but as the demand for more flamboyant colours developed, colours were developed up using two or more mixed grades of aluminium flake, making it even more difficult to match.

It is important to clearly understand that to make up an invisible repair on a metallic finish the three following factors must be correct to the original finish, and they are:

A. The colour.
B. The appearance of the film.
C. The texture and lay of the metallic.

If any of these three is not correct then the colour will appear incorrect.

47.4 The colour

This can only be altered by tinting.

47.5 The appearance

This can be altered by gun technique.

47.6 The texture

This can be altered by thinning ratios and gun technique.

Each one of the above factors will be dealt with in detail later.

Having understood the problems that lie ahead in the repair and refinishing of metallic colours, it is as well to advise foremen and management as well as insurance company assessors that more time and care is needed in the painting of damaged metallic finishes.

When the understanding of what is required is clear to all concerned then unhurried and correct progress can be well maintained.

By following carefully paint manufacturers' recommendations on thinning and spraying techniques the areas where errors may occur are drastically reduced.

The refinishing industry has been on a long and arduous learning curve for some time now and there appears to be little chance of that curve diminishing, with the spraying of pearl finishes on mass-produced vehicles during 1986.

47.7 Light reflection

When light falls onto a metallic paint film the aluminium flakes act like mirrors, and reflect the light back out. Because of this a maximum amount of reflected light will be seen when a metallic finish is viewed from the face (or at right angles to the surface). As this angle of viewing decreases so will the amount of reflected light. Viewed from the side (at an acute angle) the colour will appear darker than when viewed from the face. This difference between the face and the side tone will be most apparent when the aluminium flakes lie parallel to the surface of the paint film (Fig.25). As the orientation of the aluminium flakes becomes more random in the paint film, the face tone will darken and the difference between it and the side tone will be less. Figure 26 shows this difference and Fig.27 shows the three positions from which an observer will see the differences. This manifests itself as a 'mismatch' colour to the front wing and to the rear wing of the vehicle.

The reflection of light and the effects that metallic finishes produce are developed up further in base and clear lacquer finishes and beyond that in light emitted and light interference colours known as pearl, which are in production with major motor manufacturers.

Metallic Finishes 47.7

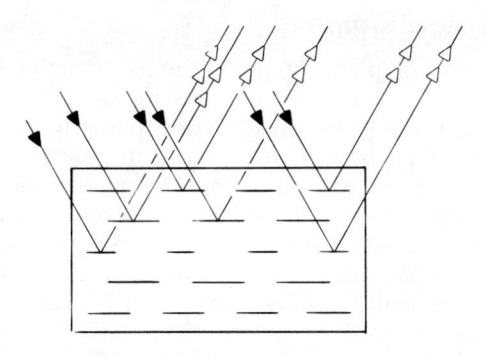

Fig. 25. Metallic light reflection

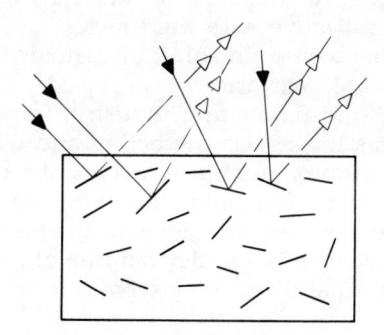

Fig. 26. Metallic reflection random

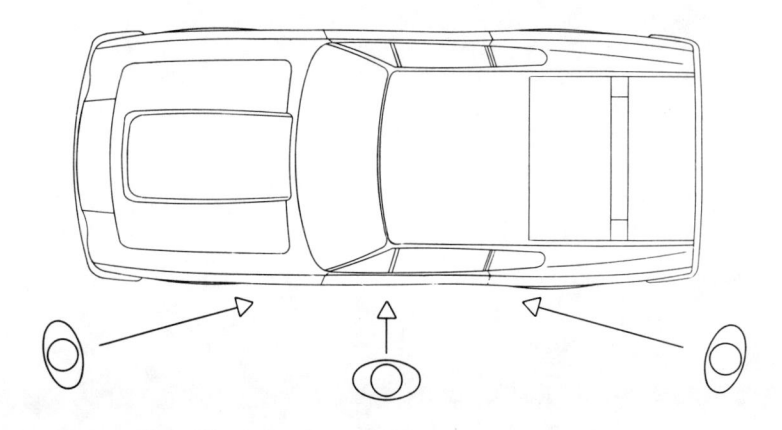

Fig. 27. Side and face tone observation

47.8 New developments

The influx of 'plastic' parts in the construction of motor vehicles, with all the interface materials necessary to repair them, also adds to the never-ending list of materials and techniques that a properly equipped and geared-up paintshop needs to ensure that all aspects of refinishing are catered for.

The level of expectancy from users will surely grow as the year 2000 draws nearer, and cars that are finished to the very highest standards, offering extended warranty on body, paint and parts, will obviously capture a higher percentage share of the market.

Similarly, refinish shops that can offer expert repairs quickly and competitively to the insurance industry and the owner will be in great demand. It is only a matter of developing skills and conditions to meet the criteria and having a willingness to understand and move with the times.

The painting and refinishing of metallics will play a major part in the whole scenario.

The sooner the refinishing industry, fragmented as it is in the UK, grasps these facts the better, and with the help of the paint manufacturers great progress will be made.

The object of a refinishing shop is to repair a vehicle in the shortest time possible and to give a proper and permanent repair to the vehicle. An understanding of metallic finishes will enhance that capability immensely.

48
Colour Match in Metallics

48.1 Matching metallics:
10 major points in the matching of metallic finishes from the metallisation of colour through to agitation in the gun pot to eliminate settle.

48.2 Background:
Colour matching is more difficult because of varying factors such as spray application on the production line and refinishing technique.

48.3 Matching original finish:
Acceptable repairs can be carried out provided the operator will adapt and attempt to match the spray techniques used on the line.

48.4 One-coat metallic:
One-coat metallic is the simple form of metallic finish where no overlay lacquer is employed.

48.5 Tinting:
Tinting with reduced strength tinters is advisable because both the side tone and the face tone are altered.

48.6 Basecoat and clear metallics:
The colours are applied in a two-coat system, consisting of colour, usually in matt form, and overlay clear lacquer.

48.1 Matching metallics

A. Metallisation of lacquer is the inclusion of minute flakes of polished aluminium.
B. The coarser the flake the better the effect.
C. Aluminium pigments are acid resistant – damp on vehicle causes mild acid.
D. When repairing film it is important to get:
   - Colour
   - Appearance
   - Texture

E. Art is in the application.
F. Vary the pressure, solvent mix and application technique before tinting.
G. Turbulence of pigment causes problems. Flotation and flocculation can occur with too wet film, which gives a darker effect.
H. Too dry a film gives a silvery light appearance.
I. It is important to test a patch on a panel.
J. Agitate the paint in the pot by stirring – aluminium settles.

48.2 Background

Colour matching is a much more complex problem than it is with solid finishes, largely because the colour obtained with both the original body colour and the repair colour is very much influenced by the actual condition of application.

Any spraying methods or conditions which produce thin coats which dry quickly will favour the orientation of the aluminium flakes parallel to the paint film surface – giving the greatest face/side tone colour difference, with a silvery face tone and a dark side tone.

Thick, slower drying coats allow the aluminium flakes more movement and therefore a more random orientation in the paint film – giving less difference between the face and side tones.

Quick drying lacquers give greater control over face and side than slower drying synthetics, because of the wider range of application viscosity and film thicknesses at which lacquer can be sprayed while still maintaining control over the aluminium appearance. Figure 28 shows a table with the different effects that conditions have on the colour.

48.3 Matching original finish

When examining some new vehicles it is possible to find overloaded, normal and lightly sprayed areas. The introduction of robotics on the line in most manufacturers has brought a new consistency to colour application, and with the use of electrostatic heads on the sprayguns the orientation of the metallic particles is greatly regularised. Nevertheless, acceptable repairs can only be carried out in the case of the former by careful attention to detail and by matching the spray techniques of the application of the original finish.

At present there are two types of metallic finish being used by the European vehicle manufacturers, but during 1986 a third finish known as pearl was introduced by Jaguar, Ford and Aston Martin Lagonda Ltd. This new light interfering/emitting finish is dealt with separately.

The current finishes are one-coat metallics, and two-coat metallics or base coat and clear.

Colour Match in Metallics 48.4

EFFECT ON COLOUR	LIGHTENS	DEEPENS
Paint shop conditions		
Temperature	Warm up	Cool down
Humidity	Low	High
Air movement	Increase	Decrease
Spraygun		
Fluid nozzle	Small	Large
Needle control	Close up	Open out
Air cap	High air consumption	Low air consumption
Fan width	Wide	Narrow
Air pressure	High	Low
Thinning		
Type of thinner	Fast	Slow
Amount of thinner	Over-thin (low viscosity)	Under-thin (raise viscosity)
Retarder	Do not use	Add up to 10% on thinner
Spray technique		
Gun distance	Distant	Close
Gun speed	Fast	Slow

Fig. 28. Table of effect of different conditions on metallic colours

48.4 One-coat metallic

These are different from solid finishes in that the colour pigment is low, and both opacity and metallic appearance are achieved by relatively large aluminium flake pigments.

After building up the required film thickness, the correct application technique is chosen for the final coat application to give the nearest match to the adjacent areas or panels on the car. The correct technique can be judged by spraying small, flexible test panels of metal or card, and when they are dry shaping them to the contour of the vehicle where the repairs are to be done.

As with colour matching of solid finishes, various techniques can be used to ease the problems. Spraying to a break or trim line will enable slight colour differences to go undetected. More effective is the technique of spraying to an edge where the angle of the panel changes.

Metallic finishes vary in lightness or darkness according to the angle of viewing. Even with the same metallic paint the colour on two adjoining panels at an angle to each other will appear different. Spraying to an edge is accomplished by careful application of two strips of masking tape. By spraying into the angle formed by the panel under repair and the projecting masking tape, a 'fade out' is obtained at the panel edge.

48.5 Tinting

Tinting may be required sometimes to match an older vehicle, or to tint a standard colour to match a 'special'.

When tinting metallic colours it is vital to remember that the tinting colours can affect either face or side tone, or both. It is therefore important to understand what colour change is required in both face and side tones. Paint manufacturers' tinting guides supply detailed information of the effects on face and side tones of their range of tinters. A simplified tinting guide is given in Fig.29.

For accurate tinting of single layer metallics, reduced strength tinters should be used.

TINTERS AFFECTING MAINLY FACE TONE	TINTERS AFFECTING MAINLY SIDE TONE	TINTERS AFFECTING FACE AND SIDE TONES
Aluminium	Oxide pigments White	Permanent blue Permanent green Permanent yellow Red dyestuff pigments Maroon dyestuff pigments Black

Fig. 29. Tinting guide for metallics

48.6 Basecoat and clear metallics

The colours are applied in a 'two-coat' or 'two-layer' process consisting of a high opacity base coat followed, wet on wet, by a clear coat. The whole is cured in one baking operation. This system gives maximum contrast between face and side 'flip' tone, combined with high gloss, a feature not always possible with single metallics.

The principles of colour matching and selection of spray techniques described for single metallics also apply to base coat and clear metallics except that the technique of adding extra clear in the fade out processes is only recommended in the case

Colour Match in Metallics 48.6

of an obvious mismatch. Adding clear will greatly change the colour of the base coat.

It must be remembered however, that:

A. The base coats are high opacity paints and full strength tinters should be used to adjust the colour.

B. The true colour is only seen after application of the clear coat, but a guide to the final colour match may be obtained by first wetting up a suitable area of the base coat with a recommended fade out thinner.

The repair of base coat and clear colours is more demanding than one-coat metallics simply because of the difference that the lacquer makes to the colour. After application of the lacquer the colour appears to 'develop' and the more lacquer coats that are applied the brighter and more intense the colour becomes. Often colour matching of base coat and clear is completed by the constant addition of lacquer coats until the final effect matches the original.

It is obvious that great care has to be taken to ensure that no dirt inclusions occur in the base coat as after the application of lacquer there is no possibility that these can be flatted out. It is also very important to observe the paint manufacturers' recommended inter-coat flash-off times. Normally it is between five and fifteen minutes in a booth. If this critical time phase is left too long then some breakdown of adhesion can occur inter-coat between the base and lacquer.

Great care must be exercised at all stages of the job, from very careful masking up to avoid bridging, which is difficult to rectify in straight colour and extremely difficult in metallic or even worse, base coat and clear.

Great care is needed in the matching and refinishing of metallic paint finishes.

49
Metallic Spray Techniques

> **49.1 Accuracy:**
> The spray application determines the appearance and texture of the metallic. The spraying must be accurate, repeatable, with a constant gunpass speed. A constant overlap percentage must be maintained.
>
> **49.2 Adjustments:**
> Test the gun on a spare 6 × 4 panel and set up the needle, air spreader and pressure.

49.1 Accuracy

When it comes to spraying metallic finishes of either one-coat or two-coat, it must be clearly understood that the accuracy of the actual application has a marked effect on the colour or appearance of the finish.

The spraygun determines the actual appearance and the texture of the finish, and the consistency and repeatability of the operator's application technique is paramount in securing a first-class job. The overlap of each pass, as well as the speed and gun distance, are now absolutely critical to the film colour.

The overlap on each pass can be as high as 80% and only after the first colour is applied can the operator be sure of the overlap necessary to obtain the correct colour, appearance and texture. This is where skills acquired in the basics are now tested to the full, and an operator must be ready to be totally flexible in his approach to the job in hand.

The basic rules still apply, and the total encapsulation of a panel in material of an even depth is what should and must be strived for. Not only skill, but also understanding and the application of that understanding, are totally necessary for a successful outcome.

Remember, the customer who signs the acceptance note will only do so if he sees a perfect repair, in other words, no visible colour or material mismatch to the original finish.

49.2 Adjustments

Test the spraygun on a panel and make adjustment to the needle, air spreader and gun speed to obtain an excellent and matching gun finish.

50
Local Repairs

> **50.1 Description:**
> Any small local repair to panel and paintwork. Normally this is private or warranty work.
>
> **50.2 Object:**
> To localise damage repair and keep cost and time to the minimum.
>
> **50.3 Procedure:**
> Clean the whole panel with compound. Fill damage area, then prime and guide coat. Make repair perfect as it is the focal point of the customer. Apply colour coats and polish out.

50.1 Description

Local repairs consist of any small repair to a panel on the vehicle. Usually these repairs are private or warranty work. Observe the damage carefully and ensure that no other minor damage exists on the panel that may have been previously overlooked.

50.2 Object

The object is to localise the spray area and keep costs in materials and time expended to the very minimum. This is particularly so with warranty work.

50.3 Procedure

First clean the whole panel and check it carefully. Cut back the surface with compound. Then:

- A. DA the damaged area.
- B. Use body filler or polyester stopper.
- C. Rub down with production paper (80 grit or finer).
- D. Wet flat and feather out with 180 wet/dry.
- E. Clean up and inspect the job.
- F. If acceptable, proceed with primer and filler coats.
- G. Allow to dry.
- H. Guide coat and wet flat with P 600 wet/dry.

50.3 Vehicle Painter's Notes

I. Clean down and inspect.
J. Spirit wipe.
K. Spray three coats of matched colour, overlapping each coat to wet out overspray.
L. Allow to dry and wet flat with P 1200 wet/dry and polish.

51
Blending Clear

> **51.1 Use:**
> When a metallic panel is part repaired and an edge to edge finish is not desirable, an application of clear lacquer in stages as a percentage of the spraying colour applied over the whole panel will allow the original finish to 'grin' through.
>
> **51.2 Method of repair:**
> Apply three coats of full colour to the damaged area.
> Add percentages of lacquer to thinned material and spray out increasing areas over the panel.

51.1 Use

Sometimes when a metallic finished vehicle is damaged in a localised area it is not necessary or even desirable to have an edge to edge finish. The development of clear blending lacquers has made it possible to paint small areas within a panel and blend the new colour away from the repaired area so that the original colour 'grins' through and makes the repair impossible to detect with the naked eye.

51.2 Method of repair

The method of repair varies from one paint manufacturer to another, but basically the same process is used universally.

A. Repair the body damage in the affected area. Ensure that this is 100% as the customer will examine the area closely.
B. Spot prime and spot fill over a well feathered area.
C. Allow to through dry.
D. Apply guide coat and block back using P 600 paper wet and dry.
E. Compound the whole panel to remove road grime from the original finish.
F. Apply at least three coats of correctly mixed and thinned colour to the spot, overlapping each application to wet down the overspray 'ring'.
G. Add 25% blending clear lacquer to the colour in the spraygun and spray the primary area.
H. Add 50% lacquer to the contents of the spraygun and spray the secondary area.

51.2 Vehicle Painter's Notes

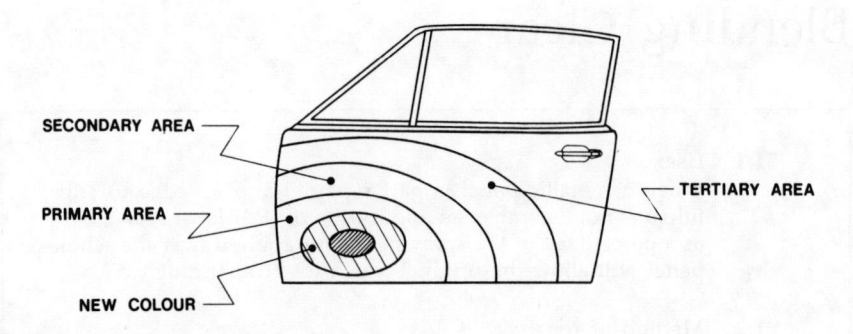

Fig. 30. Metallic blend in panel

    I. Add 75% lacquer to the contents of the spraygun and apply this over the tertiary area.
    J. Spray over the whole panel with 90% lacquer with 10% colour.
    K. Allow to through dry and then polish as necessary. See Fig.30.

52
Metallic Tinting

> **52.1 Basic rules and procedure:**
> The basic rules apply to both metallic and straight colours. Always check colours in daylight, and check the colour code on the car.
> Mix up the colour and spray out on a 6 × 4 panel, and when through dry, check against the car.
> Make all tinting alterations with care and use a very small amount of tinter at any one time.

52.1 Basic rules and procedure

When the decision has been taken to tint a metallic then a set procedure must be followed to minimise any error that may occur. The rules for colour matching are the same for metallics as for straight colours.

A. Always check colour in daylight.
B. Check the colour code on the vehicle chassis plate and cross-reference with the paint manufacturers' code book.
C. If using a weight mixing scheme, check the microfiche and make up the colour accurately to the formulation. Ensure that all basic colours are well stirred, particularly the metallic base silvers. These do tend to sink to the bottom of the can and heavy pigments need to be stirred up by hand at first.
D. Spray out correctly thinned material onto a flexible steel or cardboard primed panel.
E. Allow to through dry before offering up to the vehicle.
F. Ensure that the panel being used for the match is adjacent to the damaged panel, and that it has been cleaned back using rubbing compound.
G. Establish the need to tint.
H. Refer to the microfiche to ensure that only tinters used in the formulation act as tinters. To move away from these will cause the colour to become 'dirty'.
I. Add very small amounts of tinter and spray out each time.
J. Allow to through dry before attempting to observe the match.
K. Proceed carefully and slowly and keep the original paint back in the container so that it can be used to recover if the operator overshoots with any one tinter.
L. Do not use different silvers in the tinting.

53
Silicone Contamination

53.1 Fault:
After spraying a panel, craters appear in the film.

53.2 Causes:
Silicone contamination from car polish.

53.3 Rectification:
Drench the panel in hot water to soften and melt out the silicone. Wet flat with hot, clean, soapy water and then dry off. Use wax and grease remover. Do not use a silicone additive.

53.1 Fault

After colour spraying a panel craters appear in the film, and the colour appears to separate from itself, leaving indentations.

53.2 Causes

In silicone contamination the causes are the slip additives in certain car polishes or contamination from air hoses not designed for refinish spraying. Another cause may be airborne contamination from other processes or from nearby factory units.

53.3 Rectification

First trace the source of the contamination. Having established precisely what this is, then rectify that source. To rectify the panel, if possible wash off the colour with thinners if the material has been applied over a high stoving original finish. If this is not the case then wait until the film is dry and go through the following sequence:

A. Wet flat back the contaminated film using hot water with strong solution of liquid soap.
B. Pour very hot clean water over the panel.
C. Wet flat back with fresh hot water with liquid detergent.
D. Dry off and wipe down with spirit wipe or inter-coat wipe.
E. Spray the whole panel in filler.

F. Guide coat and wet flat back with P 600 wet and dry.
G. Colour coat in the normal way.

Do not attempt to use silicone additives to overcome the primary silicone contamination. These materials only further contaminate the spraygun, the job and the spray booth. They also tend to affect the colour surface, and poor flow results.

54
Warranty

> Warranty is the rectification of paint faults that occur in production. Often vehicles are painted at the time of the PDI, before delivery to the new owner.
>
> Motor manufacturers are looking to extended warranty times as a sales and marketing aid.

Vehicle warranty is the rectification of faults that have occurred in the manufacturing process and are recognised as such by the motor manufacturer. Often this is not fully understood by the owners of vehicles who appear to believe that a paint finish will remain without fault no matter what. Industrial fall-out, along with salt conditions from the road in winter, play a part in undermining the paint finish. If to these factors customer neglect is added, then faults in the paint film can and will occur. An owner must play his part in looking after the vehicle by washing regularly, in the proper manner, and polishing the vehicle correctly, as well as keeping the car garaged in a well ventilated garage.

The checks to ensure the quality of the paint as it arrives on the line and before it is applied to a vehicle are numerous. Both the paint manufacturer and the motor manufacturer have test labs set up to check each batch for a whole list of specifications. If the paint is below par on any one of these it is rejected and either corrected or, if necessary, discarded.

Sometimes vehicles are painted as a result of line damage in the manufacturers' assembly plant. This can be as high as 25% of vehicle production, and errors and faults can occur here as a result of application.

Warranty times claimed are examined by the manufacturers very carefully and they are indeed very keen to see that the time taken for repair falls into line with their specification for the time and material taken and used.

The end result is that the customer must have his vehicle repaired to the highest standard and as close to the original as possible.

55
Coatings

> **55.1 Coatings on new panels:**
> Important to identify which coating has been used.
>
> **55.2 Primer:**
> Some car manufacturers supply panels in primer – all ready for preparation, filling and colour.
>
> **55.3 Protective coating:**
> This looks like primer but is NOT. Ford good example – must be flatted or DA off.
>
> **55.4 Wax coating:**
> This must be removed with solvents (BMW).
>
> **55.5 Bituminous coating:**
> This must be removed with solvents – be 100% sure to remove all.

55.1 Coatings on new panels

When the bodyshop has completed a repair on a vehicle, often a wing or bonnet has been replaced with a new panel. Depending on the make of car and where this panel was supplied from, it will have a coating over it. In some cases this may well be a motor manufacturer's original primer, but in many instances this coating may be there for protection only. It is most important to identify which coating is present and take the appropriate action.

55.2 Primer

The manufacturers who supply panels already coated up in primer enable refinishers to simply carry on with the painting process, provided that no damage has occurred to the panel, either during delivery or in the bodyshop operation. The primer can be flatted and filler can be applied over the top in the normal way. Ensure that the primer is well degreased before following on with filler coats.

55.3 Protective coating

This material looks like primer but it is only a protective coating. It is usually a dark brown or black in colour and it is very dusty if flatted dry. This is not suitable as a primer as it has poor adhesion to the metal. It can be easily and quickly removed by using a chemical stripper or a DA sander.

55.4 Wax coating

It is obvious when panels are supplied in this protective wax and this must be removed with solvents. Ensure that the panel is totally clean before proceeding. (BMW use this wax coating.)

55.5 Bituminous coating

As in wax coating, the material is obvious and must be removed totally, using solvents. Ensure that it is removed completely.

56
Brushes

> **56.1 Brush application:**
> Normally used for polyesters and brushing fillers – used in commercial work for interiors of cabs and box vans.
>
> **56.2 Use:**
> Material applied by brush must be well brushed out and 'laid off' at 90° strokes – fine hair brushes will give good finish. Avoid uneven coating of material – wash brushes thoroughly after use.

56.1 Brush application

The use of brushes in the refinishing trade is generally becoming less and less common. Most brush work is carried out by commercial vehicle paintshops, but even there, the spraygun is being used more widely.

In the normal vehicle refinisher, brush work is confined to polyester application or heavy pigmented brushing fillers. The disadvantage of brushing is the brush marks that have to be blocked out after the material is dry and before spray applications can proceed.

56.2 Use

Material applied by brush must be well brushed out and 'laid off' at 90°. The finer the hair the less intrusive will be the brush marks, and therefore a good finish will be obtained. Avoid uneven coating and unnecessary build-up of material. Wash the brushes out very thoroughly after use and do not leave them dumped in a tin of solvent.

57
Wet on Wet Application

> **57.1 Definition:**
> Primer or colour coats applied immediately after application of the first coat.
>
> **57.2 Object:**
> To give a full high build of material with maximum flow out. System favoured in colours with poor opacity and metallic colours.
>
> **57.3 Application:**
> Usually a first 'set up' coat is applied then followed after flash off with the wet on wet process.

57.1 Definition

The definition of 'wet on wet' is when primer, filler or colour coats are applied one after another immediately.

57.2 Object

To give a very high build of material with maximum flow out and in the case of colour, high gloss finish from the gun. The system is favoured in colours with poor opacity and metallic colours.

57.3 Application

When this method of coating up is used it is generally applied on the two last coats of material. The paint manufacturer may call for three single coats of colour, or a single coat followed by a wet on wet, or a double header. The first single coat is for the material to 'set up' so that a heavy film build can be applied in one application.

When applying the wet on wet start at the top of the wing or door and spray on a full coat and then return to the top of the panel and repeat the procedure. If respraying a vehicle spray each panel individually in this way.

It is necessary to work quickly and accurately, as over-application can lead to sags and runs as well as solvent trapping.

Wet on Wet Application 57.3

It is important to allow a considerably extended flash off if using a low bake material. Solvent popping is a very real hazard when spraying wet on wet.

However, in air dry materials the finish that is obtained by this method is extremely good, and should be encouraged.

58
Synthetic Finish

58.1 Description:
This material is a high build synthetic with very good opacity levels. It has a gloss from the gun and is ideal for commercial work as the surface stays 'open' and will wet up overspray.

58.2 Spraying:
When spraying this material gun pass speed must increase and distance from the work piece must be greater.

58.3 Critical recoat period:
This material dries from the surface down, and therefore suffers a period when fresh applications of material will cause wrinkling.

58.4 Pressure feed:
By using pressure feed equipment the operator is able to prepare larger quantities of paint, as well as being able to keep working at the job.

58.5 Principle of pressure feed:
The principle of pressure feed is the application of low air pressure on the material in the tank.

58.6 Disadvantages:
The main problems with this material are dirt inclusions and spot repairs.

58.1 Description

This material is an alkyd synthetic high-build material that dries up in 2 to 3 hours with a critical recoat period. Full hardness takes overnight in correct ambient temperatures.

These materials are more suitable for painting commercial vehicles than lacquers that are used in vehicle refinishing, because they are 'open' longer and have a better overspray absorption time. However, the overspray will adhere on contact with a dry or nearly dry surface, it is relatively insoluble, and may be difficult or impossible to remove by polishing.

The painting sequence must be carefully planned so that all overspray falling on adjacent surfaces will be absorbed.

With the faster drying synthetics the overspray absorption time is about 10 minutes in the operating temperature (68–72° F, 20–21° C). Overspray falling on the surface painted more than 10 minutes before will remain on the surface. Overspray settling within the 10 minute period will sink into the surface and be fully absorbed.

Adequate extraction plant must be provided in the spraying area to remove all overspray from the atmosphere before the paint film has set.

58.2 Spraying

Synthetic paints are sprayed using a fluid nozzle with an internal bore no greater than 0.070 inch (1.8 mm), holding the gun 7–10 inches (17–25 cm) from the surface. The pass speed with the gun must be faster than that normally used for spraying lacquer finishes.

Low gun pressures must be avoided as they under-atomise the paint and cause sags and heavy runs.

Synthetic finishes are normally applied in two coats. The first is a light 'set up' coat or holding coat for the second and heavier application, which determines the overall film thickness, gloss, flow and general appearance of the paint film. In practice, a single pass is made and when this has 'flashed' and only a light tack remains, the second coat is applied, which is normally a double pass.

58.3 Critical recoat period

It is important to remember that synthetic finishes dry by oxidation, that is, the take up of oxygen from the air is used to polymerise the structure. The material dries from the outer surface down. This means that it suffers a 'critical recoat period'. In effect there is a period of 'wet' time after spraying is completed when the surface may be recoated and there will be no adverse effects. This is normally up to three hours. After that time, up to thirty-six hours is the period where recoating will cause lifting or wrinkling. It is important to understand this clearly and it is good practice to finish the vehicle out at one time.

58.4 Pressure feed

On commercial vehicles the remote cup spraying equipment or pressure feed units are used. This enables the refinisher to prepare larger quantities of paint ready for use. The DeVilbiss pressure feed units available will hold from 2 gallons (10 litres) to 40 gallons (200 litres) and they are the answer to large vehicle spraying.

In all cases where large quantities of the same material are to be applied, the use of the pressure feed tank is advised for the following reasons:

A. A very large amount of work can be carried out before refilling, thus obviating the waste of time that would be entailed in constantly refilling a smaller container.
B. The spraygun can be turned to any angle to coat up the work effectively.
C. The material is fed to the gun in greater volume that by any other method, particularly if heavy paints are used.
D. Less air pressure is required to obtain speed of operation.
E. Waste of paint by evaporation and other losses is eliminated.

58.5 Principle of pressure feed

The principle of pressure feed is the application of low air pressure on the material in the tank so that it is forced through a fluid hose to the spraygun. Air pressure is controlled by an air regulator on the lid, and a pressure gauge is provided. Pressure feed tanks are usually provided with a light insert container that greatly facilitates cleaning and changes of material. The tanks are strongly constructed to avoid any risk of distortion under pressure and are usually galvanised inside and outside. The lid is held on by clamps and is fitted with a gasket to prevent air leakage.

There is a safety valve, an air release valve and at least one air and one fluid draw-off cock, but on the larger tanks there are two or more air and fluid cocks, so the tank can be used by more than one operator. This is very useful if the paintshop is engaged in spraying 40 feet long box trailers.

Provision is made for a hand operated agitator to keep the material properly mixed.

58.6 Disadvantages

The main disadvantage of synthetic finishes is the problems caused by overspray or dirt inclusions. Spot repairs after dirt has been flatted out are difficult, and the ring of overspray dust as a result of this operation is very difficult to totally eliminate by polishing in the normal way. It is a case of good housekeeping and taking great care to minimise the dust and dirt problem

59
Synthetic Coach Finish

> **59.1 Type:**
> This material is a high build, high solids alkyd synthetic finish. It dries through oxidation and is usually overlaid with varnish.
>
> **59.2 Use:**
> It is mainly used on commercial vehicles and coaches.

59.1 Type

This material is a high-build, high solids alkyd synthetic finish. It dries firstly through solvent flash off to set up the surface, and secondly, through oxidation. That is the process whereby oxygen from the air is taken up into the film surface and then cross-linking or polymerisation takes place.

Due to the drying down from the exposed surface, the material does suffer a critical recoating period. This period lasts from about three hours after final colour application at 20° C, to approximately 36 hours.

59.2 Use

The material is normally used on large commercial vehicles and coaches. Often, a final coat of varnish is applied to enhance and protect the surface of the paint film.

60
Hot Spray Method

60.1 Object:
To spray pre-heated materials to a job to give better build and flow characteristics through the reduction of the viscosity level.

60.2 Low pressure spraying with heated paint
By using a paint heating system with the same spraygun and material used for cold spraying, a 20% to 30% saving can be achieved on materials.

60.3 Time saving with heated paint:
Spraying time can be almost halved by use of this method of heating paint.

60.4 The paint heater unit:
A full description of the heater unit.

60.1 Object

To spray heated materials to a job, that have been pre-heated to reduce their viscosity levels. This process enables a higher build of material with fewer coats.

Synthetic paints may be applied in a single application hot sprayed. A double pass is normally required to obtain the necessary film thickness. The viscosity of the paint is reduced by heating it to about 70° C (158° F) rather than by adding thinner.

Accurate temperature control is vital as correct application will not be achieved if the temperature is allowed to fall below 65° C (150° F). Heating above 75° C (167° F) may cause deterioration of the paint.

60.2 Low pressure spraying with heated paint

By using a paint heating system with the same spray gun and material used for cold spraying, a 20 to 30% or even greater saving can be made on the amount of paint required for a job. The method of making this saving is to thin the material exactly as for cold spraying and then to heat it to 70° C (158° F), which then reduces the viscosity still further. The paint can now be perfectly atomised with an air pressure as low as 20 to 25 psi and the fluid cut down by 20% or more, still maintaining the original film build. A short test on a panel will enable the operator to correlate the air pressure and fluid flow so that he can apply a good smooth coating. Since solvents are evaporated by the heat a smaller proportion of them reach the surface than in the case of cold spraying. Rapid cooling of the material reduces any tendency to run or sag. The paint particles hit the surface at a much lower speed, and rebound with loss of paint through spray fog is reduced to a negligible minimum.

This reduction of spray fog also gives the operator better and cleaner conditions to work in.

60.3 Time saving with heated paint

There is an alternative, major advantage in spraying heated paint which sometimes appeals to the commercial refinisher, particularly when speed is essential to meet a contract deadline. Spraying time can be almost halved simply by applying in one operation a coating twice as thick as the one normally applied. This is achieved by using the same material, but without adding any thinners at all. The paint viscosity is reduced by heat to approximately the same as it is when thinners are added for cold spraying and a normal air pressure of 50 to 70 psi is required to atomise the heated material. As before, the rapid cooling of the material as it reaches the surface reduces any tendency to run or sag.

Some degree of paint saving can also be effected by this method by adding a small amount of fast evaporating solvent so that the spraying pressure can be lowered, but the relation between the amount of thinners added and the maximum thickness of coating obtainable depends on the material used and is best discovered by experience.

Sealers, melamines, metallic paints that have stearates in them, and non gloss paints with flatting agents, can all be used in a paint heater operating at a lower temperature, at about 100° F. Little or no advantage is gained with water-based materials or spirit stains.

Figure 31 shows clearly the coating thickness obtained by hot spray methods.

60.4 The paint heater unit

A paint heating unit is equally suitable for portable and stationary use. Figure 32 shows the general layout of a stationary single operator unit, and also illustrates the method of operation of this equipment.

COLD PAINT

HEATED PAINT

SPRAYED COLD
Atomizing Air Pressure
70 p.s.i.
SOLVENT
PAINT

Coating Thickness

SPRAYED HOT
Atomizing Air Pressure
25 p.s.i.
SOLVENT
PAINT

Identical thickness to cold spray

Greater Mileage

Atomizing Air Pressure
70 p.s.i.
PAINT

Coating Thickness

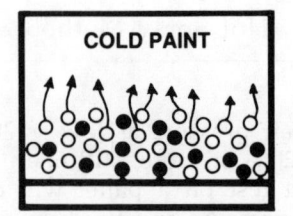

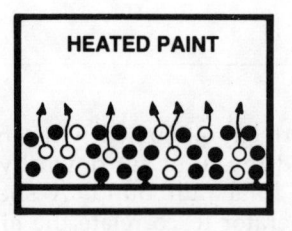

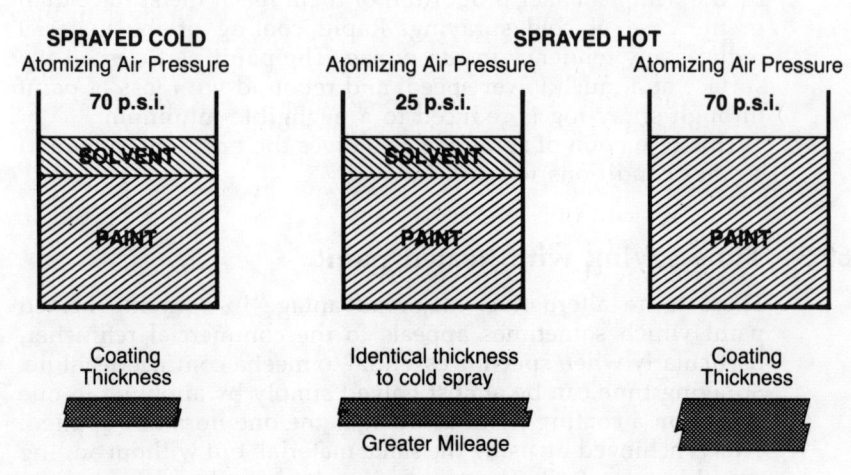

Fig. 31. Paint volume (hot and cold)

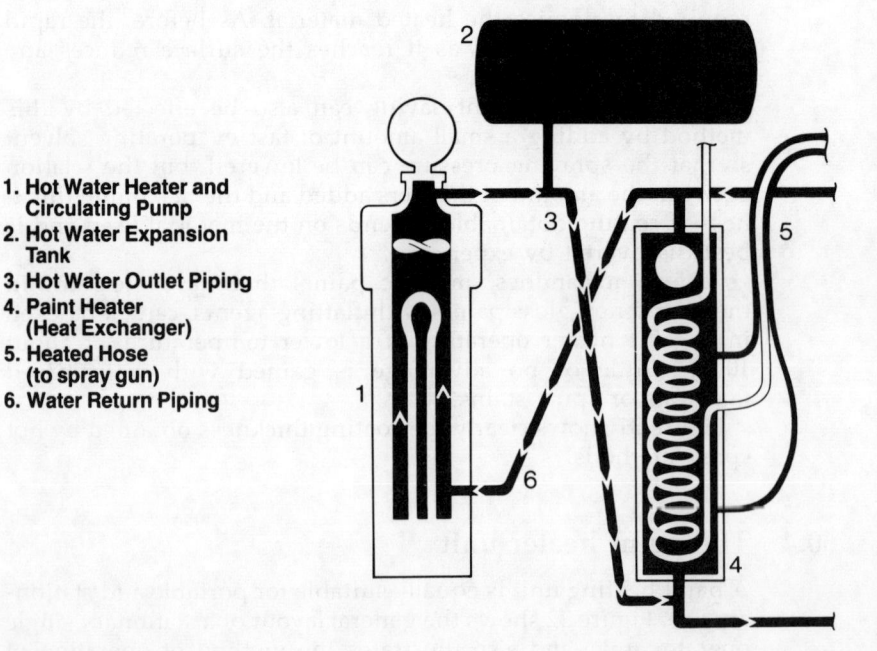

1. Hot Water Heater and Circulating Pump
2. Hot Water Expansion Tank
3. Hot Water Outlet Piping
4. Paint Heater (Heat Exchanger)
5. Heated Hose (to spray gun)
6. Water Return Piping

Fig. 32. Paint heater unit

156

Hot Spray Method 60.4

Water is pressurised to approximately 5–12 psi and recirculated in an enclosed system from a thermostatically controlled heater through a heat exchanger. Paint from a pressure feed tank or circulating system flows through tubes in the heat exchanger. The hot water, moving round the tubes in the opposite direction, transfers its heat to the paint evenly and efficiently. The paint is taken from the heat exchanger to the spraygun by means of a hot water jacketed hose which is coupled into the water system thereby ensuring that the paint is kept hot right up to the spraygun.

The use of a closed hot water system is the most efficient way of heating paint and also eliminates the fire hazard and danger of polymerisation that can occur in other types of paint heaters. For a large and busy commercial vehicle shop, hot spraying of synthetic materials is an attractive proposition and one that should always be considered most carefully.

61
Lining and Signwriting

61.1 Types:
Three types of lining on commercial vehicles: the brush method – the spray method – the transfer method.

61.2 By brush:
To the skilled signwriter the brush method is the traditional method of applying signwriting.

61.3 Gold leaf work:
A very expensive process consisting of the application of fine gold leaf to a prepared surface.

61.4 By spray:
Spray application allows the wider use of materials. More active solvent blends giving better adhesion.

61.5 By transfer:
This is the easiest method, but transfers can be very expensive.

61.1 Types

There are three methods of lining or signwriting on commercial vehicles. They are:

A. By brush
B. By spray
C. By transfer.

61.2 By brush

To the signwriter skilled in the traditional arts of the trade, brush application offers a simple process of signwriting or lining. The normal process follows this pattern:

A. Thoroughly clean the area to be painted.
B. Mark out the height of the writing using a chalked line.
C. Rough out the writing in chalk.
D. Brush the writing. Skilled signwriters will adjust the paint for both drying rate and viscosity by the addition of linseed oil and solvent. Adhesive tape or masking tape may be applied to

both the top and bottom lines of lettering to help achieve a good job.
E. Apply a second coat of colour if the opacity is poor.

Lettering in alkyd resin coach finish or one-pack polyurethane is quite durable but that done in colour-in-oil will require a top coating of varnish.

All these paints are carried in weak solvent blends, and adhesion to hard finishes such as high bake factory enamel one-pack polyurethane and two-pack arcylic finishes will not be of a high order.

61.3 Gold leaf work

A very expensive process that is very durable and still carried out on a limited scale.

The procedure is as follows:

A. Thoroughly clean the area to be gilded.
B. Mark out the height of the writing using a chalked string.
C. Rough out the writing in chalk.
D. Brush in the writing with goldsize.
E. By brush, smooth the gold leaf on to the goldsized areas.
F. Brush away surplus gold leaf.
G. Varnish the lettering or the whole panel to protect the gold leaf.

The key to success is to apply the gold leaf when the goldsize has reached the correct degree of tack.

61.4 By spray

Spray application allows the use of a wider range of paints for signwriting. Those with more active solvent blends give better adhesion to such finishes as high bake enamels, one-pack polyurethanes and two-pack acrylic finishes, but the process is long and complicated.

A typical system is:

A. Thoroughly clean area to be painted.
B. Apply a thin coat of glycerine using a finely textured sponge.
C. Press a sheet of aluminium foil onto the glycerine. Smooth out thoroughly with a plastic spreader.
D. Mark out the writing on a paper sheet.
E. Perforate the edges of the writing with a fine needle.
F. Press the paper to the aluminium foil. Fix it in place by adhesive tape at the corners.
G. Using a soft cloth dust the pin pricks with chalk when writing on dark colours, or charcoal when writing on light colours. Charcoal sticks can be easily obtained from any art shop.

- **H.** Remove the paper.
- **I.** To the pattern on the aluminium foil cut out the lettering, using a fine scalpel knife. Remove the cut-outs.
- **J.** Using an approved cleaner remove the glycerine from the exposed areas.
- **K.** Abrade any large exposed area with fine abrasive and then clean off.
- **L.** Apply the colour coats.

61.5 By transfer

This is the easiest method but transfers may be expensive depending on the number required and the complexity of the design. Three types of transfer are presently available, each having their own method of application:

- A. Self adhesive vinyl transfer.
- B. Solvent fix transfer.
- C. Goldsize transfer.

In the case of transfers it is advisable to consult carefully with the designer and the manufacturer. A company such as 3M who specialise in this type of work can give all the assistance that is required for a first-class job.

62
Transfers

62.1 Use

Commercial vehicles only – 3M make large transfers and logos to specific fleet order. When there are a number of vehicles to be affixed, technical teams demonstrate and sometimes carry out the complete operation.

Transfers are very costly and are often fixed over using lacquer or varnish.

63
Low Bake Safety Check

63.1 Procedure for safety check:
Ensure operation of oven is correct.
Ensure all flammables are removed.
Check time on and off.
Do not leave oven to run.
Check fuel in vehicle tank (one gallon in or out).
Leave handbrake off.
Do not enter oven in bake.
At end of bake – leave to cool.
Notify supervisor of any fault.
Think ahead – plan job.

63.2 Safety of waste materials:
Six points on safe disposal of waste.
Key to safety is good housekeeping and care.

The use of low bake ovens has grown dramatically in the UK and Europe. With this new aid to refinishing come a number of hazards that must be identified for safe working and the well-being of all staff and operators concerned with the day to day running of the installation.

63.1 Procedure for safety check

It is important to carry out these safety checks:

A Ensure that the operation of the oven and booth are correct. When the unit is commissioned the suppliers will take great care to see that the unit is operating correctly and safely.

B Make sure all flammable products are removed from the booth and oven prior to the commencement of the bake cycle. Any of the paint or thinners left in the oven are sure to ignite.

C Make sure that the correct time on and off is observed. Some refinishers have a small chalkboard by the oven controls so that the time is correctly recorded. This is good practice and should be encouraged.

D It is not good to leave an oven on too long because of the vehicle inside as well as the overheating that may occur in the unit itself. Ovens should not be left on and unattended, especially overnight. This is a dangerous practice.

E Check the fuel in the vehicle. The golden rule is 'either a gallon in or a gallon out'. If the tank is full then drain off a gallon and if it is empty, put a gallon in. Fuel tanks will pressurise in an oven despite the fact that they are generally placed inboard of the vehicle and slung underneath in a cooler place.

F Leave the handbrake off a vehicle in a booth in case of emergency evacuation.

G Do not enter the oven during the low bake cycle. Very hot air breathed in is very dangerous and entering an oven can cause faintness.

H At the end of the bake cycle open the doors and allow the heat to escape into the shop. Do not enter until the temperature has dropped almost to ambient.

I If in any doubt on the operation of the oven or booth ensure that the supervisor is informed immediately, and then senior management. Dangerous circumstances can arise if not attended to quickly and efficiently. The oven manufacturers have expert teams of technicians who can be brought in at very short notice to diagnose and rectify any fault.

J When using a low bake installation, try to both think and plan ahead. Good professional operators should always be watching for signs that problems may be occurring. They should responsibly advise and warn foreman and management. So often after an incident or worse, an accident, management cry, 'Why didn't someone tell us?'. It is important to act responsibly and to communicate clearly. Remember, an operator will always be thanked for stopping a potential accident, and blamed if no action has been taken.

63.2 Safety of waste materials

With health and safety at work the responsibility of everyone employed in any company it is essential that everyone acts in harmony for the total good of the company and the working environment.

The disposal of waste products is an area where care needs to be exercised.

It is important that:

A All waste products from the paint and body shop are disposed of correctly. This means the collection by specialists in solvent and paint waste disposal.

B All polyisocyanate hardeners spilled must be contained in sand and disposed of under the regulations as laid down in the 1972 Poisonous Waste Act. This material should not be allowed to enter drains.

C Mutton cloths and rags should be kept away from peroxide catalyst hardeners and in any case should not be allowed to build up.

D Waste solvent should be sealed in tins and collected by a specialist service.

63.2 Vehicle Painter's Notes

    E Empty lacquer tins should not be left open for solvents to flash off into the atmosphere.

    F Everyone should ensure that no build-up of waste materials occurs in the paint or body shop.

The key to safety is good housekeeping and care.

64
Oven probes

64.1 Specification:
In the specification of all low bake enamels a panel temperature is given at which the material will cross-link and cure.

64.2 Oven probe:
The only way to ensure that the panel temperature required is reached is to attach probes to a vehicle in the oven.

64.3 The process:
The process consists of attaching probes to the vehicle in the oven and running up the temperature and taking readings at predetemined times.

64.2 Oven run up:
The total time to reach temperature is known as the oven run up.

64.1 Specification

In the specification of all low bake materials a temperature is given at which the paint will either cross-link or force dry. This temperature refers to the panel temperature of the vehicle, and to effect a proper cure of the paint this temperature must remain steady for a predetermined period of time. All paint specifications will state, for example, 80° C for 40 minutes.

64.2 Oven probe

The only way to ensure that this temperature is reached and held is to 'probe' the oven. This is always done when the oven is first commissioned but as time progresses the oven may become unbalanced.

It is important to get an even temperature over the surface of the vehicle otherwise various faults can occur. For example, if the temperature required has not been reached then the paint in that particular area will be uncured and possibly soft. This is not to be recommended on sills, for instance. On the other hand a roof panel may be overstoved which can cause solvent popping or even edge burning.

64.3 The process

The process for oven probing is as follows:

A Probes are set on a vehicle placed in the oven.
B Probes are attached to the outer body panels.
C Probes are attached to the roof, boot, bonnet, two door panels and two sills.
D The oven is started up and every five minutes a temperature reading is taken of every probe. These are connected to a temperature gauge outside the oven.
E The readings are continually taken every five minutes until the lowest part of the car, i.e. the sills, has reached the prescribed temperature.

64.4 Oven run up

This total time is known as the oven 'run up'. This varies from oven to oven and there is no set or certain time. On average it ranges between 30 and 45 minutes.

Once the oven time has been established then the actual recommended stoving time is added to that. Therefore an oven with a 30 minute run up, using materials that stove for 30 minutes at 80° C, will require the oven total cycle to be 60 minutes.

Figure 33 shows the probe layout as used in a typical situation, and Fig.34 shows the probe report as the temperature

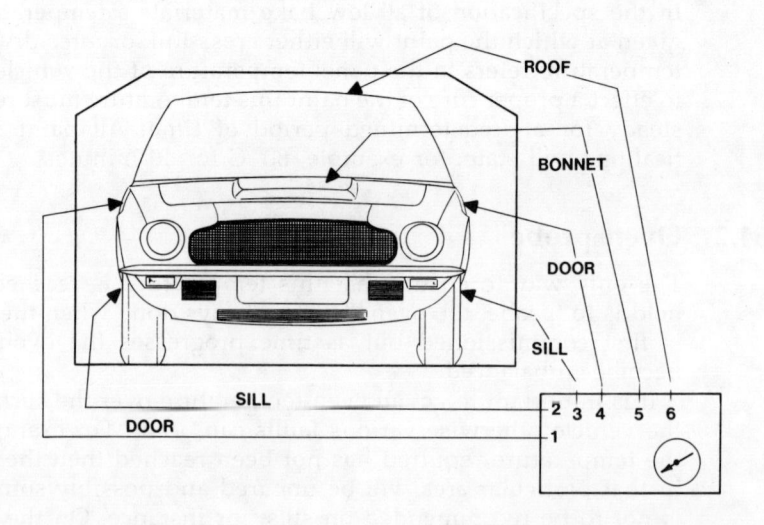

Fig. 33. Oven probe layout

Oven Probes 64.4

Time Into Oven (min.)	Probe No. 1 Bonnet	Probe No. 2 Door F/OS	Probe No. 3 Roof	Probe No. 4 Boot	Probe No. 5 Door F/NS	Diff.
0	19	18	21	21	17	6°
5	38	35	43	35	34	8°
10	50	43	58	44	42	15°
15	64	49	70	52	49	21°
20	71	54	78	59	54	24°
25	78	58	83	64	57	26°
30	84	61	88	69	61	27°
35	90	65	92	74	64	28°
40	90	65	90	74	63	27°
45	92	67	93	76	66	27°

Fig. 34. Oven probe result sheet

rises. Sometimes the oven probe can show and identify when a hot spot is occurring. Oven balance is difficult, and they need to be set up by the oven technicians who know precisely where to pinpoint oven temperature faults.

If, for instance, a refinisher is suffering with solvent boil on the same area of every car, let us say the back of the roof or the front, depending on which way the vehicle is facing in the booth, it would appear that a hot spot has occurred. Use of the probe will identify that and the experts can be called in to rectify the situation.

Oven maintenance is important and most users enter into a contract with the oven supplier to service the unit on a regular basis.

65
Low Bake Ovens

> **65.1 Object:**
> The object of having a low bake oven is to enable the refinisher to match as closely as possible the finish obtained by the motor manufacturer.
>
> **65.2 Ovens:**
> There are two types of oven installation, the combined spray booth and oven known as the combi, and the separate low bake oven.

65.1 Object

The object of having low bake installations was so that the refinisher could match the original finish used by the motor manufacturer more closely and speed up the turnround of work, with the added bonus of improved quality.

There are two types of oven. There is the combined spraybooth and oven known as the combi, or the separate spraybooth with the oven set adjacent to the spraying facility.

The oven unit gives a rapid system in which the paint film is cured and hard by the time the vehicle has cooled to ambient temperature from the oven.

The combined booth and oven units have become more popular due to the fact that they use less floor space, and the fact that minimum baking temperatures have come down from 80° C to 60–70° C (140–160° F). Warm up time required for the unit to reach stoving, i.e. the run up temperature following the changeover from spray cycle to bake cycle is thus reduced.

65.2 Ovens

The two types already described give different advantages. The separate booth and oven give much greater output, and the combi unit is cheaper to install and requires less floor space.

Heating may be by oil, waste oil, gas or electricity, and it must be said that oil and gas fired ovens are the most popular. These ovens are normally heated indirectly, the fuel being burned in an enclosed heat exchanger. In electric installations heating is by infra-red units mounted inside the oven.

Low Bake Ovens 65.2

The air flow principles for oil, gas and infra-red electric ovens are the same as those for spraybooths, i.e. the air flows from the ceiling to the floor. Whereas the spraybooths have an open circulation system with fresh air being drawn in all the time and passed through the booth and exhausted, stoving ovens have a closed system, the air being continually recirculated. To give efficient, quick, economical heating this air must be circulated at a minimum velocity of 100 linear feet (3.3 metres) per minute over the vehicle body. If the velocity is too high, oven popping or small blistering is likely to occur.

In a combined unit the overall area of airflow may have to be decreased during the stoving operation, so that the necessary velocity over the body is achieved.

To prevent a build-up of explosive mixture, 10% of the air is continually discharged to the atmosphere. This is usually achieved by placing an exhaust pipe of about 5 inches diameter in the roof of the oven. Pressurisation and natural convection currents normally produce the necessary discharge through the pipe. The air loss may be made good by replacement air introduced through the filters on the vacuum side of the air impelling pump. The recirculated air is filtered prior to reheating.

For force drying, metal temperatures up to 160° F (70° C) are used, whilst for true low bake where full cross-linking of the paint molecules takes place, a temperature of 80°C 176° F is required.

Figure 35 shows a typical low bake oven circulation lay out.

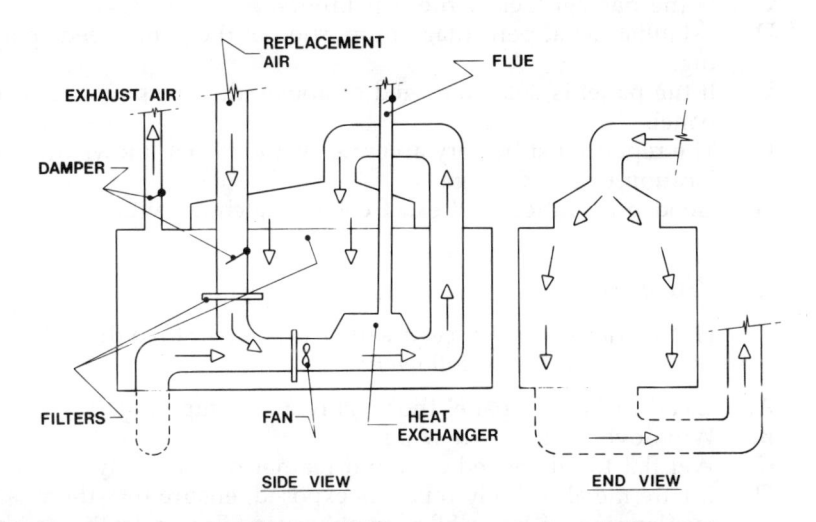

Fig. 35. Oven layout

66
Local Repairs (Straight Colours)

66.1 Key points:
There are a number of key points to consider before attempting a local repair in straight colour. For example, what percentage proportion of any given panel requires painting.

66.2 Procedure:
A full process and procedure for carrying out local repairs.

Often, with light paint or panel damage, the refinisher will be faced with a decision as to whether to paint the complete panel or attempt a local repair.

66.1 Key points

There are different techniques for straight and for metallic colours, but before proceeding with any local repair the following key points must be considered:

- **A.** The amount of damage – the deciding factor.
- **B.** Where is the damage?
- **C.** Is the damage below the waistline?
- **D.** Establish what percentage proportion of the panel needs painting.
- **E.** If the panel is 30% damaged or above, then respray the whole panel.
- **F.** The repair must be very accurate, as the owner knows the exact location of the damage.
- **G.** Good preparation and exact colour match are vital.

66.2 Procedure

Having decided to proceed with a local repair then the following sequence should be followed:

- **A.** Clean the whole panel thoroughly and compound.
- **B.** Wipe over with spirit wipe.
- **C.** Wet flat the damaged area and feather out carefully.
- **D.** If bare metal or body repair is exposed, ensure that there is no overlapping of body filler or polyester filler on to the original paint finish. See Fig.36.
- **E.** Brush on self etch primer to exposed metal.

Local Repairs (Straight Colours) 66.2

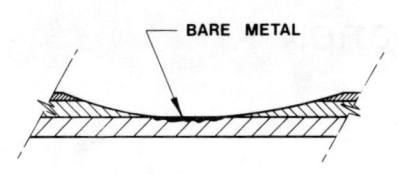

Fig. 36. Bodyfiller repair

F. Spot fill with primer filler.
G. Cut down the flow rate and air pressure from the gun to give more accurate control.
H. Build up the surface with three or four filler coats, overlapping with each pass to take up the feather edging (Fig.37).

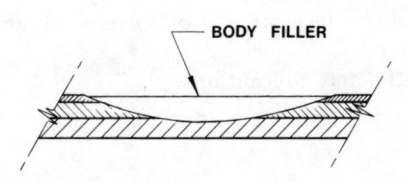

Fig. 37. Filler overlapping paint

I. Allow to through dry.
J. Apply a thin black guide coat.
K. Block repair using P 600 wet and dry paper.
L. Clean off and leather dry.
M. Examine the surface very carefully and look for sinkage.
N. Wet flat over the filler and surrounding original paint with P 1200. Extend the flatting to about 3 to 4 inches onto the original finish.
O. Mix and match the colour to the vehicle. Ensure an exact match.
P. Apply three coats of colour, overlapping each one to wet up the overspray edge.
Q. On the final coat of colour, thin the paint in the pot of the gun to 10% paint and 90% thinners and lightly dust onto the edge of the repair. Proceed until a 'wet out' has occurred and the overspray is dissolved down.
R. Allow to through dry and then compound and polish.

The key points in the operation are:

- Careful preparation.
- An exact colour match.

If this painting operation is carried out correctly it can be very successful, and saves time and money. Fewer materials are consumed and flatting and preparation time is cut to a very minimum.

67
Polyurethane

> **67.1 Type:**
> Normally two-pack type – will not harden until catalyst added. Material is two-pack polyurethane – two-pack epoxy primers and finishes. Polyester stoppers and spray fillers.
>
> **67.2 Disadvantages:**
> Short pot life – essential to clean guns well immediately after spraying.
>
> **67.3 Advantages:**
> Low solvent level or free of solvent – high build and solid contents.
> **Take all safety precautions.**

67.1 Type

Polyurethane is a reaction product of either alkyd resin, polyester resin, or acrylic resin fixed with an isocyanate. It is a product that has been used for many years in the commercial painting field, as it has many advantages. The types of material are:

A Urethane alkyds, used in one-pack polyurethane finishes, the resin components being already reacted and no free isocyanate is present.

B Urethane polyester and urethane acrylic, the modern 2K refinish paints which are two-pack finishes, now becoming widely used in the UK refinish industry. In this material the paint and the isocyanate hardener are mixed just prior to use. Free isocyanate is present and fresh air safety breathing apparatus should be used when spraying this material in a controlled spraybooth with proper extraction.

C Urethane elastomers. These are specially flexible resins for plastics. The components are already reacted and no free isocyanate is present.

67.2 Disadvantages

The disadvantages of this material are:

A A short pot life in the case of two-pack materials.

B   Dirt inclusions in the one-pack materials due to a longer set up time.
C   The need for proper safety equipment.

67.3 Advantages

The advantages of this material are:

A. High build and opacity levels.
B. Excellent gloss from the gun.
C. Excellent colour retention.
D. Excellent weatherability.
E. Ease of application.
F. Can be low baked or air dried.
G. Can be put into service immediately after through dry.

    The development of this material over the last decade has been exceptional and the growth in its sale and use by the refinish industry has been marked. The improved materials that will evolve from this firm foundation hold out interesting prospects for the future.

68
High Gloss Finish in Cellulose

> **68.1 Preparation:**
> The key to the whole concept lies here. The preparation of the vehicle must be both detailed and accurate.
>
> **68.2 Procedure for original finish:**
> The full procedure for painting over a sound original substrate.
>
> **68.3 Procedure from bare metal:**
> The full procedure after stripping old paintwork back to the metal.
>
> **68.4 Excellent results**
> To achieve excellent results time and care must be expended.

Sometimes a refinisher may be called upon to produce a very high gloss finish on a vehicle. Normally this request is linked with specialist vehicles of some kind, e.g. an old Bentley or Jaguar, or perhaps the latest, and most advanced sports GT of some kind. However, owners will insist that nothing but the best will be suitable, and this demand will be reflected in the charge made for the work.

The process that follows is for nitrocellulose finish.

68.1 Preparation

The key to the whole concept lies here. The preparation of the vehicle must be detailed as well as accurate.

Depending on the condition of the vehicle, either wet flat if sound or strip to bare metal if in doubt.

68.2 Procedure for original finish

A. Wash the vehicle thoroughly and dry off.
B. Blow out all channels and returns.
C. Wet flat with P 600 after attending to minor body repairs.
D. Wash vehicle down again and blow out channels.
E. Check the surface of the flatted paintwork carefully.
F. Mask up.
G. Spirit wipe.
H. Spray two or three coats of primer filler as per the manufacturers' specification.

I. Either air dry overnight or low bake for 15 minutes at 80° C.
J. Guide coat.
K. Wet flat with P 600 wet and dry, using a block on any repairs and on flat panels.
L. Wash down to remove all sludge, and dry off.
M. Blow out all channels and returns.
N. Remove masking and clean up edges.
O. Apply cellulose stopper if necessary.
P. If stopper is used, then block down when dry, and spray a coat of cellulose primer filler over the top to seal it back.
Q. Spirit wipe.
R. Re-mask the vehicle.
S. Spray three coats of cellulose colour correctly thinned with 30 minutes flash off between coats.
T. Allow to through dry overnight or low bake for 30 minutes at 80° C.
U. Wet flat the whole vehicle with P 800 wet and dry, blocking flat panels and repairs that may show. Dry off and spirit wipe.
V. Apply two coats of cellulose colour correctly thinned with 30 minute flash off between coats.
W. Allow overnight to dry or low bake as outlined above.
X. At this point either P 1200 wet and dry the vehicle and then compound and polish, or wet flat with P 800 and spray a single coat of colour followed by a wet on wet, allow to air dry through, and then compound and polish the result.
Y. De-mask and polish the finish to a high gloss with a cutting polish.

68.3 Procedure from bare metal

A. Ensure the metal is absolutely clean and free from any contamination.
B. Mask up.
C. Lightly abrade the surface with P 320 wet and dry.
D. Dry off and spirit wipe.
E. Apply a single coat of self etch primer.
F. Apply either a cellulose filler or epoxy filler.
G. Allow to completely dry through and then guide coat.
H. Wet flat using a block where necessary, with P 600 wet and dry.
I. Dry off and examine the surface carefully for imperfections.
J. Stop up with cellulose stopper if necessary.
K. Seal back any stopper with a coat of cellulose primer filler.
L. Proceed as from Q. in procedure for original finish.

68.4 Excellent results

To achieve excellent results many careful and patient hours must be expended. There is no quick way to refinish a vehicle to a very high standard without the input of time.

68.4 Vehicle Painter's Notes

It is good practice to examine the vehicle at every stage, and make adjustments to the surface to ensure perfect panel shape. Any imperfections will certainly show when high gloss finish is applied.

Cellulose can be polished to a very high gloss and this operation should be carried out with care, so that the clarity of the colour may show. A 'milky' appearance along with fine scratches can be removed with cutting polishes, such as T Cut, and final wax polishing should take place several weeks later, to allow the film time to fully harden. If wax is applied too soon it can dull the finish.

If the finish is properly cared for it will last for many years without difficulty.

69
Low Gloss

> **69.1 Causes of low gloss:**
> A full list of causes including low temperature, high humidity and over-atomisation of the material.
>
> **69.2 Rectification**
> The fault can be easily rectified by compounding and polishing.
>
> **69.3 Oven haze:**
> This is a temperature fault that causes a haze on the surface of the film.
>
> **69.4 Rectification of oven haze:**
> This can easily be overcome by compounding and polishing.

69.1 Causes of low gloss

It is always more desirable to complete a refinishing job with a good gloss obtained from the gun. A fault that often occurs is a low gloss final finish. Other than poor application by the operator there are several causes, and these are as follows:

A. Too low a temperature in the booth.
B. A condition of high humidity.
C. The use of cheap or unsuitable thinners.
D. The primer filler coats too absorbent.
E. The over-atomisation of paint.
F. Poor air circulation and extraction level too low.
G. The material is faulty.

69.2 Rectification

The fault can be easily rectified by compounding and then polishing the panel. In severe cases, however, a flat back with P 800 wet and dry followed by repsraying the panel. It is important to rectify the root cause of this problem rather than follow the path of constant rectification.

69.3 Oven haze

In low bake operations oven haze can occur. This shows as a low gloss, but is due to the oven operation, and the causes are:

A. Temperature fault in the oven cycle causing over-stoving.
B. The ventilation in the oven whilst on stove is poor or unbalanced.
C. A fault in the oven itself.

69.4 Rectification of oven haze

Compound and polish or in severe cases wet flat and refinish.

70
Safety Forum

> Safety the duty of every employee – safety committee – report any dangerous conditions – particular hazards in refinishing industry.

Under the Health and Safety at Work Act, it is the duty of every person employed to ensure that all safety requirements are met and maintained. It is the duty of everyone to be responsible for his own safety and that of his fellow workers. It is not a management responsibility only and this must be clearly understood by everyone.

A safety committee should be established and the members of the committee should meet on a regular basis to discuss safety within the company or organisation.

Individual persons should approach members of the committee to make known any dangerous conditions or malpractices.

It is a good principle to have a general discussion group within the workforce at regular periods so that there can be an open and frank discussion regarding safety.

In the repair and refinishing of vehicles there are a number of hazards that need to be clearly identified and all the necessary safety precautions taken.

Chapter 1 has already outlined the hazards within the paint shop but there are others which apply to the bodyshop, such as welding and the use of GRP (glass reinforced polyester).

The importance of health and safety at work cannot be overstressed and the more serious discussion followed by constructive action to prevent any accident, the better.

Never hesitate to point out dangers and bad practice to everyone concerned. Not to do so is an abdication of duty.

71
Polyisocyanate Finishes

> **71.1 Description:**
> A high solids urethane with lacquer drying speed through the use of an isocyanate hardener.
>
> **71.2 Surface preparation:**
> The surface preparation is minimal. Either wet flat or dry flat the panel and use cellulose primer fillers.
>
> **71.3 Safety:**
> It is important that all safety procedures are followed carefully when using this material.

71.1 Description

The material is a high solids urethane with lacquer drying speed through the use of a catalyst, which is polyisocyanate. The benefits of using this material in a busy main agent or refinishing shop are many. The excellent build and opacity level make the part repair of damaged vehicles easier, and give a hard finish that is very durable, and very similar to the vehicle manufacturers' original finish. When low baking this material it saves time because of the little time required to cure. 15 minutes at 80° C panel temperature is a great improvement on standard low bake requiring 40 minutes at that temperature.

The material can be air dried for refinishers that do not operate a low bake oven.

71.2 Surface preparation

The material requires a minimum of preparation due to the high solids and opacity, and it can overlay any other finish without causing any film disturbance, provided that the original film is sound. The only two restrictions that the material has are:

A The limited pot life once the catalyst has been added to the colour.
B The safety procedures that must be observed when using the material.

71.3 Safety

It is important that:

- A The vehicle to be sprayed is earthed.
- B The spray operation takes place in the booth and **never** in the open shop.
- C The booth has a proper air change rate.
- D Where possible the refinisher must spray away from himself.
- E The refinisher should use gloves and barrier cream.
- F The refinisher should use air line breathing apparatus.
- G No other person should enter the booth whilst spraying is taking place, and at the end of the spray operation the refinisher should remain in the booth for at least three minutes to allow the spray mist to be fully evacuated from the booth.
- H All operators using this material should have a medical before the introduction of the system.
- I Any spillages of polyisocyanate should be absorbed into sand.
- J These spillages should be disposed of as under the 1972 Poisonous Waste Act. It must not be allowed to enter drains.

These high-build materials are excellent and their use will spread even more in the UK, but the safety when using them is of paramount importance to the refinisher.

72
Material Types

> Lacquers – acrylic lacquers – synthetics – two-pack materials: the different properties compared.

The properties of the different types of material currently on offer are as follows:

72.1 Lacquers

- Nitrocellulose lacquer has a rapid surface dry. The material dries by solvent evaporation.
- Dirt inclusions are not a problem as they can be easily flatted out.
- The solid content is low (28% to 32%).
- There is a high paint usage.
- It has good scratch filling properties.
- The thinners are strong and can wrinkle old substrates.
- The material is very polishable after full through dry.
- It is most widely used by smaller refinishers throughout the UK.

72.2 Acrylic lacquers

- These lacquers dry faster than nitrocellulose.
- They give excellent gloss and colour retention.
- It is vital to have carried out accurate and fine preparation.
- The thinners are a strong solvent and will cause wrinkling on old unstable substrates.
- Should not be sprayed over any other finish except original motor manufacturers' OE.

72.3 Synthetics

- These dry out by oxidation after the initial 'set up' by solvent evaporation.
- They dry from the top surface down.
- They have a high percentage of solids (50% to 52%), giving very good opacity or covering power.
- The thinner is basically white spirit and therefore weak. This means that synthetic can be coated over any other finish.

- The material does not need polishing. It offers good gloss from the gun.
- It holds 'open' for some while allowing large vehicles to be painted without overspray dust problems.
- The film is soft and easily marked.
- It has a critical recoat period of between 3 hours to 36 hours after application of final colour.

72.4 Two-pack materials

- These materials are becoming more popular in the UK.
- They are high build acrylic urethanes with the through dry of a lacquer by the use of a catalyst hardener.
- The hardener is normally polyisocyanate.
- Safety precautions must be observed when spraying.
- All spraying must be carried out in the spray booth.
- The material can air dry, be force dried or low baked.
- It is very resistant to all types of damage and airborne contaminates.
- It is easily repaired.
- It can be polished.
- Dirt inclusions are not a major problem.
- The material can be sprayed over all other finishes providing that they are sound.
- This is the repair material of the future.

73
Electropriming

> **73.1 Background:**
> In 1963 can lacquer developed for food industry – application by deposition.
> Electropainting is deposition of paint film from an aqueous solution (colloidal dispersion in water) by means of a direct electrical current passed through the surface to be coated to the paint.
>
> **73.2 Advantages:**
> Safe – uniform film thickness – no runs – low cost.
>
> **73.3 Disadvantages:**
> Only single coat of paint – high cost of plant.
> Ford Halewood – tank holds 20,000 gallons – 35 bodies per hour, 1 body in under 2 minutes.

73.1 Background

Electro-deposition is a comparatively new technique which has shown considerable development in the industrial painting field. It has many advantages as a means of coating metal, and is particularly effective with water-based priming paint.

Electro-deposition achieves better coverage of sharp edges and also of difficult inaccessible internal surfaces, for example those occurring in intricate mouldings used in the fittings for car bodies.

The ability to find a way to these hidden surfaces is spoken of as the 'throwing power' of a paint.

Further advantages of the method are the better control of the thickness of the coating, a reduced wastage of material, the avoidance of sagging and drips, and a shorter drying and stoving schedule.

The process requires a bath of paint into which the article to be coated can be immersed and subjected to the action of an electric current passed between the article and an electrode also immersed in the bath. In aqueous paints a voltage of as little as 100 V can give the desired coating in a minute or so.

The mechanisms involved in the process are complex and due to the migration of charged ions and particles under the influence of the applied potential difference.

Most motor manufacturers are now using electropriming as the base for the finish.

73.2 Advantages

It is a safe process giving a uniform film thickness, and depending on voltage and immersion time this can range from ½ thou (12.5 microns) to 1½ thou (37.5 microns).

73.3 Disadvantages

The process can only apply a single coat of material.
The cost of the plant and equipment is very high, and needs careful maintenance and control.

74
Vehicle Washing and Care of Film

> Importance of good after-care.
> Wash in warm water – corrosive salt and grime in winter months – thorough removal of soap – leather dry – waxing.

This is a most important aspect of a fine finished vehicle. The after-care of a finish will give the lasting service that an owner should demand.

Always wash the vehicle in warm water with the addition of a liquid soap, about a teaspoon to a gallon or 5 litre bucket. This is mild enough to remove road film as a wetting agent without damaging the paint film itself. The vehicle should always be wetted up with warm water before attempting to sponge it down. A great deal of damage is done by people washing mud and grit off with a sponge on a dry surface and it is almost the same as wet flatting the vehicle. Thoroughly clean the car with several applications, especially in winter months when the salt and grime is so corrosive. Wash your car down as often as possible in this way. After this operation hose the car with a low pressure cold water hose to remove all traces of the liquid soap. Hose under the car as much as possible to break down the salt.

Leather the vehicle dry with a good quality chamois, taking care to rinse out often in warm clean water.

A paint film needs waxing two or three times a year with a good quality wax. Before this operation it is important to wash and then to clean off the road film by using T Cut. It is most important that this is done as it is unlikely that washing in warm soapy water will disperse the road film completely.

After T Cut apply two coats of wax and polish to a high lustre.

Wash your vehicle often, it improves the film. As you will observe, chauffeur-driven limousines are always immaculate due to the continuous washing and leathering.

75
Airless Spray

75.1 Description:
It is a process whereby paint is sprayed without the use of compressed air.

75.2 Use:
The paint is pumped into a special gun at high pressure, up to 3,000 psi, and as the paint crosses the orifice of the gun the fall to atmospheric pressure causes the paint to break up into a fine spray.

The use of airless spray is normally confined to commercial vehicle refinishers. It is widely used in the marine and civil engineering industries.

75.1 Description

It is a process of spraying paint fully atomised without the use of compressed air.

75.2 Use

The paint is pumped into a special gun at varying pressures up to 3,000 psi and then when released produces a fine spray. The paint can be hot sprayed at a lower pressure range, i.e. 400 to 1200 psi.

As the paint and solvents cross the orifice of the gun, they vaporise due to the drop in pressure from high value to atmospheric. The solvent boil that follows produces volume expansion and automatic disruption of the paint stream into fine globules.

When this process is used it causes electrostatic potential and the gun must be earthed.

There is no gun control as such, just off and on.

Using this material and process ensures that there is no 'fog' or overspray, no rebound nor any wastage.

The equipment does need extensive cleaning after the operation.

76
Safety Reminders

- Pay attention to your job and actions.
- Keep the workshop clean and tidy.
- Do not smoke in the paintshop.
- Do not weld in the paintshop.
- Wash hands and ensure the removal of all paint before leaving.
- Keep inorganic peroxides safe.
- Do not enter the oven when it is on.
- Always push vehicles around the workshop.
- Do not eat in the paintshop.
- Always use protective clothing and masks.
- Ensure that all waste cloth is disposed of correctly.
- Ensure that waste solvent is sealed in tins and collected.
- It is the duty of every person employed in a workshop to ensure his own safety and that of his fellow worker.
- Take every measure possible to avoid accidents.
- Report all accidents when they occur.

77
Assessments of Time and Materials

77.1 Time and materials:
With cost increasing, management is forced more and more to look closely at these areas to maintain profitability.

77.2 Productive work:
Work which alters the physical or chemical nature of the product.

77.3 Ancillary work:
Service or any other work related to a process which is not classified as productive.

77.4 Lost time:
Lost or diverted time where an operator may be obtaining tools, etc.

77.5 Waiting time:
Part of attendance time waiting for materials, for example.

77.6 Excess work:
Extra work because of re-work.

77.7 Qualified worker:
The definition of a qualified worker.

77.8 Standard rating BS34016:
Standard rating over the working day or shift.

77.9 Rating scales BS34015:
Standard Rating Scale.

77.10 Time:
Time taken by a qualified man to complete a particular job.

77.11 Materials:
Materials that are used within the job.

77.1–77.5 Vehicle Painter's Notes

77.1 Time and materials

As labour times and material costs increase, management have had to look more closely at these areas to maintain profitability. Times taken within the paintshop must be within the times estimated to complete the work. Although estimates given to insurance companies and private customers are 'estimates' they nevertheless have to be of reasonable accuracy. Full understanding of the time necessary to complete a job that will enable the refinisher to obtain a signed satisfaction note, and be sure that the work is of a quality that will suffer normal wear and tear over a period of warranty, is necessary for sensible estimating.

It is obvious that a first-class refinisher can produce excellent work, but if the estimate and final bill only equal the labour and material costs then a no-profit business will fail.

It is necessary for refinishers to remain competitive, and this should be done by staffing with professionals and using the latest body repair and painting equipment available. The initial cost of this is high but a sound investment for the future.

In work study the basic definitions are as follows.

77.2 Productive work

Work which alters the physical or chemical nature of the product or advances the process as a necessary contribution to its completion. For example, rubbing down primer filler or the application of colour coats.

77.3 Ancillary work

Service or any other work related to a process, which is not appropriate to classify as productive. For example, obtaining tools, cleaning surfaces.

77.4 Lost time (diverted time)

That part of attendance time during which the worker is engaged on other than productive or ancillary work. For example, searching for tools, attending meetings or rectification of equipment.

77.5 Waiting time

The part of attendance time during which the worker is available but is prevented from working. For example, waiting for a special colour to be delivered.

77.6 Excess work

Extra work occasioned by departure from the specified method or materials for which control standards have been established. For example, re-work, polishing out dirt inclusions.

77.7 Qualified worker

A qualified worker is defined as a person who:

A. Is physically capable,
B. Has the required intelligence,
C. Has the necessary education, and
D. Has acquired the necessary skills and knowledge, to carry out the work to satisfactory standards of safety, quantity and quality.

To help identify more clearly the time taken by a worker, there is a British Standard rating and scale.

77.8 Standard rating, BS34016

'The rating corresponding to the average rate at which qualified workers will naturally work at a job, provided they adhere to the specified method and provided they are motivated to apply themselves to their work. If the standard rating is maintained and the appropriate relaxation is taken, a worker will achieve standard performance over the working day or shift.'

77.9 British Standard Rating and Performance Scales, BS34015

The British Standard Scale is 0–100 where 0 corresponds to no activity and 100 corresponds to standard rating. Figure 38 shows the British Standard Scale.

77.10 Time

Time taken by one qualified man in the paintshop should be carefully monitored so that a true understanding of what exactly happens when preparing and painting a vehicle is gained, and the estimate for the job relates to that.

The BIA at Thatcham have completed an enormous programme of in-depth study and the manuals that they publish should be used as the guide for every repairer. For example, the BIA state that in order to make allowance for standard operations undertaken on every job, a further addition, called

77.10 Vehicle Painter's Notes

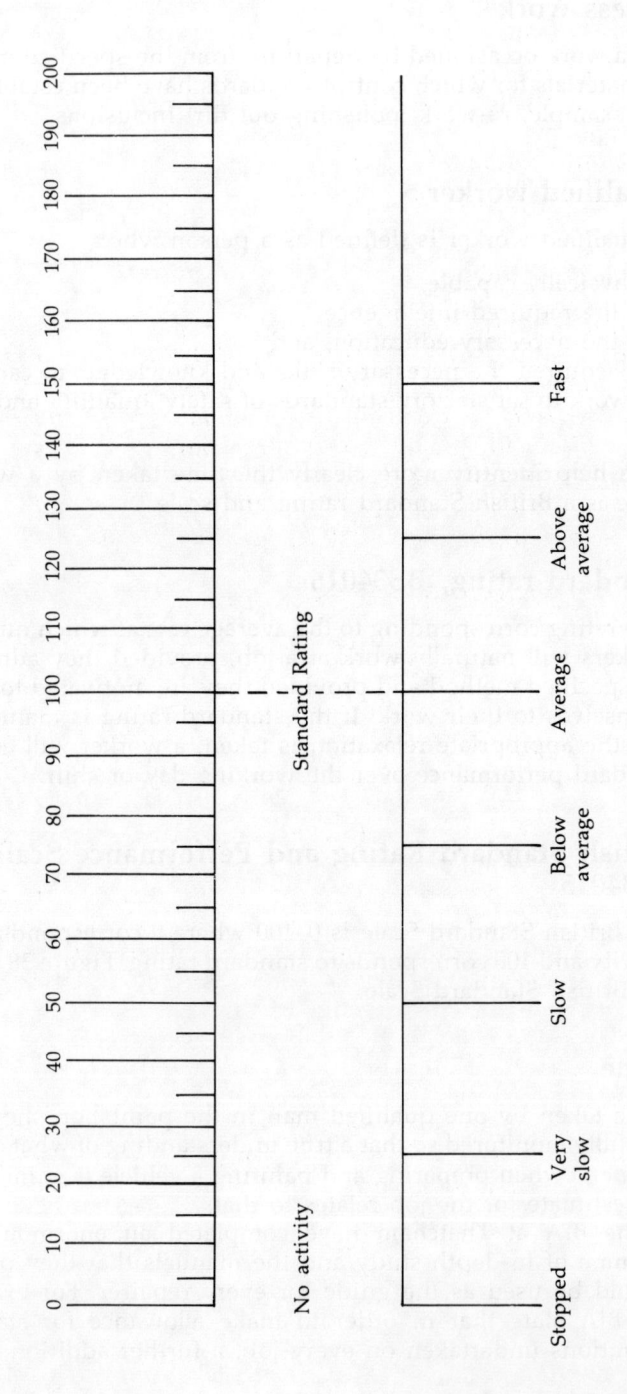

Fig. 38. British Standard scale of work

- 200
- 190
- 180
- 170
- 160
- 150 — Fast
- 140
- 130
- 125 — Above average
- 120
- 110
- 100 — Average — Standard Rating
- 90
- 80
- 75 — Below average
- 70
- 60
- 50 — Slow
- 40
- 30
- 20 — Very slow
- 10
- 0 — Stopped

No activity

192

the Job Allowance, is made to the total repair time, to cater for the following:

Vehicle movement in and out of the workshop	6 mins
Obtaining tools outside the immediate work location	10 mins
Obtaining parts and materials from stores	9 mins
Recording time on job card and time sheet	5 mins
	30 mins

Further to this, it is important to be realistic about time taken to mask up or flat down a panel correctly. These operations are very time-consuming and they need to be carried out by qualified operators. A clean-up time should also be added to the paint time and should be carefully monitored.

When management, by careful study, can identify the time taken to complete a particular job to a quality standard that is totally acceptable, then the time taken can relate exactly to the estimate.

In the interest of profit, invest in professional operators and first-class equipment.

77.11 Materials

Materials used on a job can be carefully itemised as they are used at each stage of the operation. Qualified staff will be able to monitor this very carefully and the correct charge for every sheet of wet and dry and every litre of paint can be correctly attributed to that particular job.

The VBRA does publish regularly the current prices, trade and retail, of all items used in a paintshop. This ranges from a single sheet of flatting paper to a litre of unthinned colour. These prices are accepted by insurance company assessors and are not normally challenged.

(See Fig.23, Chapter 28, for the table of paint use drawn up by the BIA.)

The use of a weight mixing scheme in the paintshop does ensure that the over-ordering of colour does not occur. As the colour mix is fully repeatable then any small additional amount may be quickly produced.

78
Estimating

> Highly skilled operation – analysis of job procedure and time taken to achieve finish. Examine damage and paint film carefully – check general condition of vehicle for standard – ensure price/hours are plus – profit and loss determined by estimate. Make careful notes on every aspect of repair – written estimates must be well detailed. Supplementary estimate as more damage revealed. Talk to customer about vehicle and accident.

Estimating is a highly skilled operation and one that so often goes wrong. Normally in a repairer's, one person gives the estimate. Usually this person is a motor engineer, which is the way it should be for the body and chassis damage, but when it comes to the painting of the damaged vehicle then an engineer leaves a lot to be desired. The BIA at Thatcham quickly realised that insurance assessors knew little or nothing about paint and did include this topic within the training that they offered to the insurance industry. The painting estimate should be carried out by a qualified painter, who will observe the condition of the vehicle and whether it has been repaired before. So often problems arise in the painting system due to previous repairs, often poorly carried out.

As the paint finishes on OE have become more and more sophisticated, the understanding has to be that much greater. With the majors moving onto pearl colours, the need for a complete and total understanding of what is required will intensify still further.

Management should now be looking at the estimate being prepared by the reception engineer and the paint foreman. It may be necessary, for instance, to paint the complete side of a vehicle in pearl if the colour build, which will affect the colour density, is difficult to capture as an edge to edge. In other words to refinish the front wing only on a pearl finish will inevitably lead the customer to reject the vehicle due to a mismatch of colour. Although that may not be the case, it will be impossible to persuade the owner. Commercially, it is a fact that the vehicle should be refinished once and once only, and if the customer refuses the work then any futher time spent on the car is a total dead loss.

For some general pointers for consideration it is important to:

Estimating 78.1

A. Check the general condition of the damaged vehicle. The customer's expectancy plays a major part in deciding how the painting operation will be carried out.
B. Ensure that the hours quoted are on the plus side to cover any difficulties that may arise. With paintwork the refinisher is very much alone, with only his own skill to carry the job through.
C. Make careful notes on every aspect of the repair. Leave no observation to memory, write it all down.
D. Written estimates must be well prepared and very informative.
E. Do not hesitate to call in the insurance assessor, or owner in the case of private work, to discuss further damage that may be revealed during the strip down and repair of the vehicle.
F. Prepare supplementary estimates with the same care and attention to detail as the primary or main estimate.
G. Take time and care, as the estimate is the future profit of the company.
H. Beware of the owner who wants you to 'flash a drop of colour over the bonnet while you've got the gun in your hand'. Extra work must be paid for at the prescribed rate.
I. For customer goodwill, talk to the owner about his accident, and as it's never his fault he will want to tell you. A happy and satisfied customer is the very best type.
J. Use the BIA manuals for sensible times as a base line.

79
Acrylic Finish

> **79.1 Type:**
> A blend of resin from the acrylic family which includes perspex and PVC (poly vinyl chloride) and synthetic plasticisers.
>
> **79.2 Characteristics:**
> The material has a rapid surface dry and excellent gloss.
>
> **79.3 Application:**
> The surface preparation is critical and must be sound. The correct thinners and viscosity must be used.

79.1 Type

Acrylic finishes are made up from a blend of the acrylic family which includes perspex and PVC (poly vinyl chloride) with synthetic plasticisers.

The material is on a water white resin, methyl methacrylate, and is impervious to discoloration by ultra violet light. Because of this, colours, particularly white, remain constant and will not 'yellow' or discolour over long periods of time. Exposure on the 5-year Florida test shows virtually no movement in colour deterioration.

79.2 Characteristics

The material has a rapid surface dry with excellent gloss.

The colour retention is first class and the material is very polishable after full through dry. In air dry this is 16 hours or overnight, or force dry after up to 1 hour at a panel temperature of up to 80° C. The material dries by solvent evaporation.

79.3 Application

The surface preparation is critical. It must be both sound and finely flatted. Paper should be no coarser than P 600 wet/dry. Any scratch marks will show in the final film.

A. Prepare correctly.
B. Use the correct thinner, and thin to the manufacturers' recommendation.

C. Ensure the panel temperature is to the shop ambient, which should be 68–72° F or 20–21° C.
D. Ensure the paint to be sprayed is also to ambient temperature, otherwise 'cobwebbing' can occur.
E. Where the original finish is a sand bake sand process, such as G.M.Vauxhall, it can help to spray a single coat and wait up to 20 minutes before continuing. This will allow the new solvent to gently penetrate the original film which has a certain film tension due to the reflow process.

80
Rolls-Royce

> Original cars sold were chassis and engine only – Coachbuilders Vanden Plas – Thrupp & Maberly – Barclays – built bodies on for customers – chassis took road shock – allowed coachpainters to coat panel work to excess. Rolls-Royce produced own bodies after World War Two – introduction of Shadow in 1965 – monocoque structure – paint systems modified – epoxy filler used followed by acrylic. Cars polished – vehicles hand-worked for results obtained.

In finishing and refinishing, whenever the ultimate is being discussed it is always described as 'a Rolls-Royce finish'. The standard and quality that Rolls-Royce have been able to set and maintain is in itself a legend. The finish on the cars is almost the ultimate, being surpassed by only one other company, Aston Martin Lagonda Ltd, who, because of their very low build figures, four cars a week, are able to devote even more time to paint and panel work than Rolls-Royce.

Both concerns use a TPA material over a sound substrate of epoxy primer filler, with great care and attention at every stage of the operation.

From 1906 to 1939 Rolls-Royce produced chassis only, and it was left to specialist coachbuilders to construct coachwork to the individual requirements of customers. After the Second World War it was decided to produce a complete car with Rolls-Royce becoming responsible for the coachwork of the newly introduced steel saloons. To do this required more extensive factory space for paintshop and assembly areas and so the motor car production line was re-established at Crewe in a factory that had built aero engines throughout the war. The Mulliner Park Ward Division of Rolls-Royce Motors maintains the traditional crafts and skills of English coachbuilding.

To refinish a Rolls-Royce that has been damaged is not difficult but great care must be taken at every stage, and the skill required by the operator revolves round the discipline that he can input to the job. Preparation is of paramount importance, and careful examination of the panel work, the primer, the filler coats and then the colour at every stage will ensure a fine finish. It is advisable to use the Rolls-Royce material that can be supplied and follow the instructions for use carefully.

Care of the vehicle is important and the parts of the vehicle that are not to be painted should be covered with a dust sheet. It is necessary to mask up for primer application and after wet

Rolls-Royce 80.1

flatting clean down and leather dry and de-mask. Then clean up edges where primer overspray dust may have collected. Clean out all channels and then re-mask for colour application.

It will be necessary to wet flat with P 1200 and compound and polish to achieve the lustre of the original paintwork.

If care is taken throughout then the work can be finished in one operation to a very high standard.

Plan the work carefully and the owner will not be disappointed.

81
Spraying Wheels

81.1 Procedure

When spraying wheels the following procedure should be followed:

A. Remove the wheel from the vehicle.
B. Thoroughly wash with hot soapy water.
C. Dry and degrease with spirit wipe.
D. Wet flat with P 180 wet/dry.
E. Feather out rust spots.
F. If the tyre has not been removed then mask up.
G. Degrease with spirit wipe.
H. Spray a thin coat of primer.
I. Spray two coats of filler.
J. Work round the wheel using low pressure from the gun.
K. Allow the material to 'fall' into the wheel pressings.
L. Allow to dry through.
M. Dry flat with P 600 wet/dry paper.
N. Spray wheel silver or body colour into the wheel.
O. Always use low pressure and overthin the material slightly so that the colour will 'wet up'.
P. Allow to fully through dry before refitting to the vehicle.

82
Matt Black Cellulose

82.1 Type

A nitrocellulose black with a matting agent added at proportions that will give different reflective surface.

82.2 Mix

Use unthinned material for mixing. The following proportions will give the stated result after spraying using a fast thinner.

A. 75% black to 25% matt agent thinned for 21 to 24 seconds at 20° C will give a semi-gloss finish.
B. 50% black to 50% matt agent thinned for 21 to 24 seconds BS4 at 20° C will give a semi-matt finish.
C. 25% black to 75% matt agent thinned for 21 to 24 seconds BS4 at 20° C will give a flat matt finish.

82.3 Application

Over suitably primed or well flatted original finish. Three single coats are usually sufficient to obtain the desired result.

83
Paintshop Relations

> **83.1 Object:**
> Good relationship, work as team, need help often, e.g. lining – removal of bonnet – fitting bumper. Result, fast turn round of work – more bonus.
> Important to work closely with bodyshop – no 'them and us' – overall consideration is customer satisfaction.
> Be accurate in work – need for team work. Be enthusiastic about good work standard.

The dividing line between bodyshop and paintshop has always been ill-defined. Exactly where the bodyshop finishes and the paintshop begins has been a point of contention between the two since the repair and painting of carriages commenced.

Body men believe that the paint hides every blemish and misshapen panel, and the paint men are convinced that body men are totally unaware of the limitations of paint. As usual, the truth lies somewhere in between.

83.1 Object

It is important to foster good relations and make staff understand that it is similar to the blacksmith and the saddler, it's all on the horse, which is owned by the paying customer. The trades are indeed different and require different skills and understanding. When two shops work well and in harmony then profits can rise dramatically, and if a bonus scheme is in operation, take-home pay can rise considerably.

It is essential that the body and paint men cross the divide to improve work, or commence at an early stage. For example, the paintshop spray in the back of a wing before the door is refitted, or the body shop ease out little 'dings' in a new door skin which have become apparent whilst rubbing down primer filler, and which therefore will not need slow-drying cellulose stopper.

Working together will promote a good atmosphere and ensure that the company is profitable, and stays in business, and the wages of the staff continually improve.

Be enthusiastic about good quality work, it is infectious and even the customers become enthusiastic.

84
Powder Coatings

84.1 Background:
Because of the technical strides made in the technology surrounding powder coatings the motor manufacturers are now using these materials.

84.2 Advantages:
The advantages of using powder coating are many, but the main advantages are excellent mechanical properties, use in automatic plants and fire and explosion risk greatly reduced.

84.3 Manufacture:
The control is accurately maintained during the manufacture of powder coating.

84.4 Preparation:
Preparing the parts to be coated is as important as the application. They must be dry, free from grease and totally clean.

84.5 Solvent cleaning:
This is a simple method of cleaning but is subject to misuse.

84.6 Alkali cleaning:
A wide range of alkali cleaners will remove most oils and greases.

84.7 Ultrasonic cleaning:
This system can be used in conjunction with other cleaning methods.

84.8 Process:
The stage-by-stage process of finishing using powder coating.

84.1 Background

In recent years great strides have been made in powder coating technology and many motor manufacturers are now using these materials.

The history of powder development and Arthur Holden & Sons Ltd's involvement go hand in hand. The company commenced investigation into the processes of powder coating in 1963 and its commitment has continued ever since with considerable investment in both development and manufacture.

The first powders, which were based on epoxy resin, were applied at high air pressure to ensure a full cloud; film thicknesses were around 100–125 microns and stoving was based on 200° C for 30 minutes.

Considerable progress has been made since then with both powders and equipment. Standard materials can now be applied with film builds as low as 35 microns and cured at 160° C article temperature for 10 minutes. Powder emission rates are much lower and transfer efficiency greater, so far less overspray has to be collected and re-cycled.

Resin manufacturers have played their part in the development of powder coating. Resins and curative systems have become more sophisticated and new resin types have become available. To a considerable extent the mixed polymer system epoxy polyester has replaced epoxy as the general purpose powder.

Neither epoxy polyester nor epoxy can be recommended for outdoor applications when chalking cannot be tolerated, although the protection afforded by the film will be unimpaired. For outdoor durability polyester is normally recommended.

84.2 Advantages

Certainly suppliers of small parts to the motor industry are coating up with powder because of the advantages, and it is worth noting them:

- Powder usage normally implies a one-coat system. For specialised purposes, occasionally it may form part or the whole of a two-coat system.
- Overspray may be collected, sieved and recycled for further use.
- Powder is used straight from the container in which it is supplied – no viscosity checks, thinning, meter readings, as are necessary with wet paint.
- Mechanical properties are excellent, resulting in far less damage when piece parts are handled and assembled.
- Powder coating may be machined.
- Powder is easy to apply by hand application, an unskilled operator soon establishing the correct technique.
- Powders lend themselves to use through automatic spray equipment with ease of control of film build.
- Fire and explosion risks are reduced as compared with solvent-borne paints – see Chapter 1 on Health and Safety.
- Virtually no volatiles are involved, and therefore less ventilation is required.

- No flash off zone is required between spray section and stoving plant – hence saving in space.
- Powder plants, particularly automatic plants, occupy less space than wet paint systems.
- Properly designed powder application systems offer considerable energy saving compared with most wet paint systems.
- Correctly formulated powders have good edge coverage and do not 'run'.
- Plant maintenance is easy. Basic requirement to clean recovery booth is a suitable vacuum cleaner.

84.3 Manufacture

The actual manufacture of powder coating is carefully controlled, and resin, pigments, hardeners and additives such as flow agent are loaded into a blending machine which mixes the raw materials for a determined time to obtain a homogeneous mix. Samples are taken from the dry blend mix, extruded on small scale equipment and checked for colour (using a colour computer), gloss and appearance against the master pattern. If incorrect, the necessary additions are made, the blend is remixed and further checks made. It is imperative that the mix be correct at this stage as, once extruded, no further adjustment can be made. When correct, the dry blend mix passes on to the next stage – extrusion.

The dry blend mix is fed to the extruder, the barrel of which is heated to a temperature allowing the mix to become molten. The screws of the extruder, when revolving, cause 'shear' forces to be exerted on the premix which disperse the pigments and additives into the molten resin system. The hot mixture emerges from the die head on the extruder to be passed through a cooling system, usually water-cooled rollers, followed by further cooling on a conveyor.

Extrusion temperature varies from 80° C to 130° C depending on the type of system being manufactured, and extrusion conditions are carefully controlled to ensure that no premature reaction takes place. Cooled extrudate, which is brittle at ambient temperature, passes through a kibbler (or crushing machine) to produce pieces approximately $\frac{1}{4}$ inch square.

Kibbled extrudate is passed through a grinding machine to produce a finely ground powder of a predetermined particle size distribution. Granulometry can be adjusted to suit the requirements of a particular application equipment.

The final operation of sieving removes any oversize particles.

At all stages of the manufacturing process cleanliness is important to avoid contamination, whether of one colour by another or between non-compatible resin sytems. Cleaning is a time-consuming operation which can be minimised by careful production planning, but not eliminated, and downtime remains a major item of production expense. Since downtime is the same for small and large batches, a minimum batch size has to be stipulated.

84.4 Preparation

Preparing the parts for coating is as important as the application. The requirements for successful powder coating are that powder be applied to clean, dry, grease-free material. A coating which appears well adhered initially may break down after a very short time if the substrate pre-treatment was inadequate.

The presence of oils and grease, whether deliberately applied as a temporary protection or carried over from a previous engineering operation, can be readily detected, so too can mill scale and rust. More troublesome, because they are both less apparent and more difficult to remove, are mould lubricants on castings and die lubricants on extruded sections. The surface extruded aluminium sections have, in addition, a film containing magnesium oxide which must be removed.

In addition to cleaning, positive pre-treatment may be required to enhance adhesion and increase long-term corrosion protection. Chemical pre-treatment of ferrous and chromate treatment of aluminium substrates inhibit underfilm corrosion creep and a well applied pre-treatment/powder coating system will give satisfactory performance for many years.

Extreme examples are bus chassis, for which epoxy powder on zinc phosphated hot-dipped galvanised steel is specified, and aluminium window extrusions, which use a system of polyester powder over pre-treatment.

84.5 Solvent cleaning

This is a simple method of cleaning, but prone to misuse. Dirty rags and dirty solvents can put back almost as much contamination as they take off. Clean rags plus solvent used for a limited period then discarded must be the order of the day.

Vapour degreasing using trichloroethane (which is replacing trichlorethulene on health and safety grounds) is a much favoured system to remove oils and grease, but it will not remove soils adequately and a further wipe over with a clean cloth may be necessary. This disadvantage may be overcome by utilising the materials in a pressure jet system.

84.6 Alkali cleaning

A wide range of alkali cleaners are available to remove most oils, greases and soils even up to heavy concentrations. Special grain refining agents can be included to promote specific phosphate coating properties. The workpiece may be either dipped or sprayed with the hot aqueous cleaner but must be thoroughly rinsed, preferably twice, before passing on to the next stage of the pre-treatment.

A system utilising sulphuric and/or hydrochloric acid will remove rust and light scale. The workpiece must be rinsed thoroughly before the next stage of the pre-treatment.

84.7 Ultrasonic cleaning

Ultrasonic agitation can be allied with various cleaning methods to assist in the removal of soils.

This system will remove rust and millscale but incorrect choice of shot can result in a surface profile on the substrate which is unsuitable for powder coating. If powder is to be applied at 50 microns, profiles should not be above 25–30 microns. G07 and G12 grits have been found to be suitable.

After shotblasting, surface dust must be removed prior to the next stage which must take place without delay, as freshly shot blasted surfaces rapidly oxidise.

84.8 Process

After thorough cleaning, the workpiece should pass immediately to the phosphating stage and the norm is an orderly progression through a series of tanks or spray stations. Different phosphate systems are available for ferrous, zinc coated, and mixed ferrous/aluminium systems, and the advice of the pre-treatment supplier should be taken on the one most suitable for the application.

In recent years low temperature phosphating systems as outlined above have been introduced with process temperature reduced from 80–90° C to 35–40° C as an energy saving measure.

Duraplast powder coatings are formulated for application by electrostatic spray. Application may be by a single hand gun or one or more guns forming an automatic set-up.

Powder from the feed hopper is conveyed on an airstream to the gun where it is given an electrostatic charge. The charged particles are attracted to the earthed workpiece. Powder adheres through electrostatic attraction and is difficult to remove by normal air movement or jolting. (Figure 39 shows a typical set-up for powder application.)

84.8 Vehicle Painter's Notes

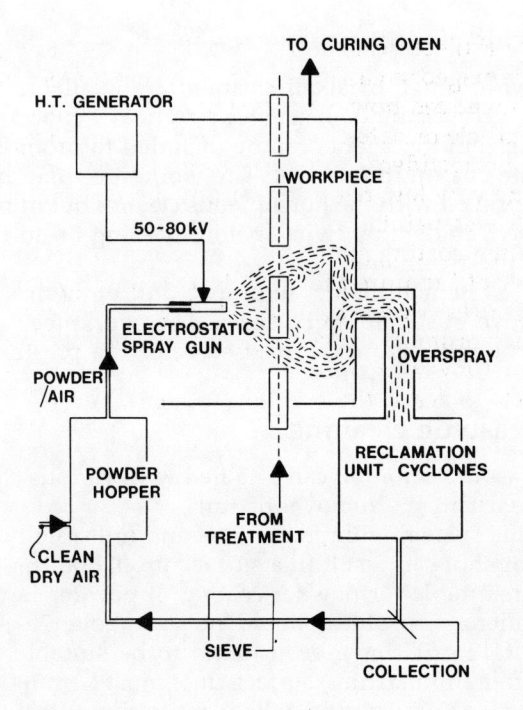

Fig. 39. Powder coating layout

Powder emission rates, air pressures and voltage settings are normally established at the plant commissioning stage in close collaboration with equipment and powder suppliers. Operating experience and changing work loads may necessitate adjustments, but constant 'fiddling' with the controls should not normally be involved.

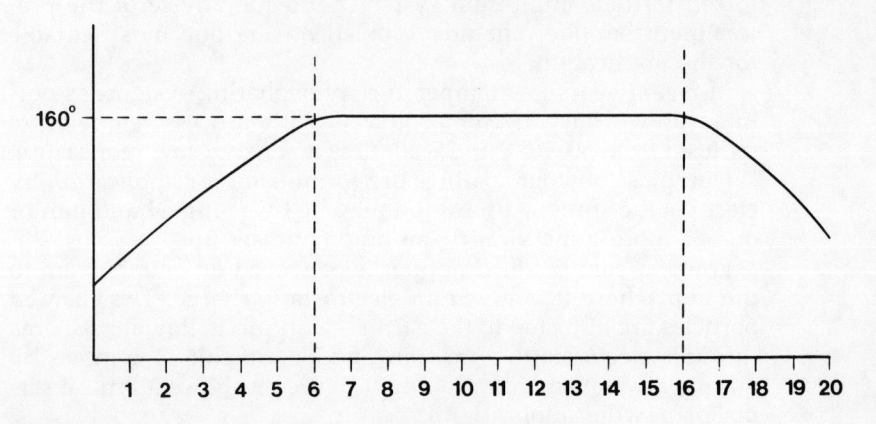

Fig. 40. Power coating temperature profile

Workpieces can be inspected prior to stoving and any bare or damaged areas touched in with powder. In the last resort, the unstoved powder can be blown off with compressed air and the article recoated. The ease and convenience with which faults can be rectified at an early stage is a factor which contributes to the very low reject rate of powder coating finishes compared with wet paint.

After coating, the article is moved directly to an oven (no 'flash off' required for powder) where the powder melts, flows and cures. Recommended cure schedules are normally quoted as, for example, 160° C article temperature for 10 min (see Fig. 40). (Allowance must be made for the time necessary for the article to reach the specified temperature.)

85
Pearl Finishes

85.1 Background:
A new development in OE has been the introduction of pearl effect colours. The appearance of this finish is striking and will without doubt be the future trend in styling.

85.2 Description of optics:
In pearl effect pigments there are two basic requirements: one, the flake or shape, and two, a high refractive index.

85.3 Three-coat system:
The three-coat system consists of a base white, the pearl, and a clear overlay lacquer.

85.4 Repair:
General comments and guidelines for the repair of the material.

85.1 Background

A further new development that has taken place recently is the introduction of 'pearl' finishes on the production line at major motor maufacturers.

It is estimated that during 1986–87 up to 30% of all original finishes will be 'pearl' effect. Without doubt these finishes are very attractive indeed and will certainly help boost sales of new cars. For the manufacturer, there is no undue problem in the application of such colours on the line. However, the problems commence when repairs are necessary. It is not impossible to repair pearl finish but certainly the process demands higher skills and a complete and total understanding of the whole procedure.

The paint manufacturers are able in a two-coat system, clear over base, to incorporate the light interference and light emitting mica into the metallic base coat. This gives such a side effect as to be able to virtually change the colour when viewed obliquely and over contoured panels gives a striking appearance.

The light reflectivity is governed by the film weight of the base and therefore the more colour applied the deeper it will appear. Only after the application of the lacquer with the isocyanate hardener will the colour be fully developed, thereby leaving an area of uncertainty as to the colour match edge to edge. It is likely that future metallic finishes will all be of pearl effect and the refinishing industry will have to rise to meet the challenge that the repair of such finishes will bring. Competent and well-trained operators will have to be on hand to cope with this development.

All of the paint manufacturers are running training schools and they will be in a position to train personnel in the use and repair of these finishes.

85.2 Description of optics

In pearlescent pigments there are two basic requirements. One, a flake or plate-like shape and two, a high refractive index. The plate-like shape comes from a very thin layer of titanium dioxide or iron oxide deposited on the flat mica surfaces.

Pearl pigments are quite transparent, and this can be an advantage and a disadvantage.

A The disadvantage: being transparent, pearl pigments have poor opacity and will not cover or hide well.
B The advantage: being transparent they do not 'muddy up' or distort the colour value by being 'dirty' in the film, or the organic colourants used in conjunction with the pearl pigments that achieve the polychromatic effect.

Many pearlescent pigments look like a metallic, especially the iron oxide-coated mica type, but are really not metallic. Many are capable of showing colour through the light interference effect. This is caused by controlling the total thickness of the titanium dioxide layer on the mica. As the thickness of the coating increases, light interference occurs in an optical phenomenon similar to that observed in soap bubbles. These light interference colours have a two-tone effect; the stronger colour is seen by reflection and the weaker colour by the transmission of light.

When interference colour type of pearl pigments are added to transparent organic colours, the two-colour interplay is known as goniochromaticity. This means that the resultant colour can look very different at different viewing angles. This makes the refinishing of such colours difficult.

Another optical effect is that when pearl pigments are used in the make-up of colour, there is a greater difference in the 'flop', to the extent that a totally different colour is observed. For example a pearl white may change to a green or blue when viewed from the reverse or 'flop'.

85.3 Three-coat system

A recent development in the pearl finish is to use the material in a three-coat system. This consists of a ground coat or undercoat in white or black, which must be totally unbroken on edges, etc., followed by the pearl colour with its poor opacity and then a clear overlay lacquer. Usually a two-pack isocyanate type of material.

The application of the pearl in this system has to be very accurate otherwise a 'blotchy' appearance can result.

85.4 Repair

This is where the refinishing industry will change gear to meet the challenge of this new type of material. Only by careful and methodical working practice will it be possible to begin to repair vehicles painted in this material as original equipment.

Firstly, the amount of paint damage to be rectified must be carefully noted. If more than one-third of a panel or collection of panels, or side of the vehicle is damaged, then the whole panel, collection or side must be refinished. The difficulty comes in the matching, not only of the face tone but the side tone or 'flop', due to the vast differences already alluded to.

In the three-coat system the light intensity or reflective value of the colour will be related to its opacity level, which will vary as the colour coats are applied. Simply put, the more colour coats of pearl the darker and deeper the colour will appear under the lacquer overlay.

The only way to match this will be by spraying out a large separate panel which has been contoured to produce a shape equivalent to the vehicle panel. As each colour coat is applied a careful note must be kept of the application levels. It is possible and most probable that complete resprays will be the quickest and most cost-effective way of refinishing these vehicles. Knowledge and the application of that knowledge will be the only way that the refinishing industry will cope with this new styling material.

86
Painting Composites (Plastic Parts)

> **86.1 Background:**
> The use of composites or plastic parts will increase substantially over the next ten years.
>
> **86.2 Types:**
> The two groups presently being used, the flexible and the hard plastic.
>
> **86.3 Refinishing:**
> Choice of refinishing system of the components depends on the composition of the material.

86.1 Background

Some designers feel that steels and light alloy castings will face very tough competition in the future from thermoplastics. Now, according to a British Plastics Processing Working Party Report, *British Plastics, The Next Ten Years*, the most obvious growth of plastics will occur in the construction of automobiles. In 1980 the average European passenger car used 55 kg of thermoplastics. By 1990, it will be over 98 kg per car. From the automobile's consumption of around 20% of the world's output of plastics, this figure is anticipated to rise to over 50% by the year 2000.

To refinish these plastic or 'composite' parts a careful process of preparation must be followed.

Correct pre-treatment, e.g. removal of the parting agent, and correct surface preparation followed by a suitable primer, are necessary to obtain good adhesion.

Some plastics are sensitive to solvent attack and may soften or craze. In such special cases isolating primer must be used.

86.2 Types

Plastics may be classified into two groups:

A Flexible plastics for which specially plasticised flexible paints are necessary.
B Hard plastics.

86.2 Vehicle Painter's Notes

PLASTIC	COMMENTS	MAX. STOVING TEMP.	PREPARATION	PRIMER	FINISHES
GRP glass reinforced polyester	Pinholes may be present	60°C	Remove release agent and clean. Flat with P 400 and spirit wipe. Fill blow holes.	Standard	Standard
PC polycarbonate	Solvent sensitive	60°C	Clean. Flat with P 600.	Special isolating primer	Standard
ABS acrylonitrile butadiene styrene	Solvent sensitive	60°C	Clean thoroughly. Do not flat.	Special isolating primer	Standard
PPO polyphenylene oxide (noryl)	Sensitive to strong solvents	60°C	Clean thoroughly. Flat P 600.	None	Standard
PP/EPDN polypropylene ethylene propylene diene rubber	Poor adhesion	60°C	Clean thoroughly. Flat P 400/P 600.	Special adhesion primer	Requires flexible additive
PU polyurethane	Removal of release agent difficult	60°C	Clean thoroughly. Flat P 600.	Special flexible primer	Requires flexible additive
PA polyamide (nylon)		80°C	Clean thoroughly. Flat P 600.	Special flexible	Standard
PBT pocanpolybutylterephthalate	Fairly flexible	60°C	Clean thoroughly. Flat to remove imperfections. Use P 800.	Special flexible primer	Requires flexible additive

Fig. 41. Table for painting plastics

86.3 Refinishing

Some hard plastics may be affected by the painting operation and their resistance to shock substantially weakened. Such plastics also require special flexible finishes.

The motor industry is moving towards the use of plastic alloys suitable for production line painting. The choice of refinish system for such components will depend on the composition of the alloy. Figure 41 shows the table for refinishing plastic.

87
Electrostatic Spraying

> **87.1 Background:**
> More popular in industrial applications but becoming more widely used by refinishers.
>
> **87.2 Advantages:**
> Some of the many advantages are greater saving of materials and its suitability for a wide range of paints.
>
> **87.3 Equipment:**
> A list of equipment that can be supplied by DeVilbiss Company Ltd.

87.1 Background

Electrostatic spraying has become more popular in recent years both in industrial environments and with refinishers. It is a method by which overspray is much reduced, employing an electrostatic field to direct and confine the spray deposition to the surface of the object being coated. The spraygun is connected to a source of very high voltage and the article to be coated is earthed so that the spray emerging from the gun is immediately drawn to the object. A further advantage of electrostatic spraying is that parts which are not easily accessible, sharp edges, such as car doors, and intricately constructed objects are more certain to receive a coating. The process gives a 'wrap round' effect and on car doors, for instance, the overspray that passes the edge is drawn back to it from the other side.

When used with metallic finishes the electrostatic application gives a complete and uniform metallic lay and is therefore both correct and attractive in appearance.

87.2 Advantages

The advantages of using electrostatic spray equipment are:

A. It gives greater material savings achieved by external charging of the paint.
B. It is suitable for use with a wide range of standard coatings.
C. It is reliable, easy to maintain and service, and simple to use.

D. Maximum operating efficiency ensures a first class finish on localised work or in high production.
E. Reduced spraybooth maintenance due to higher transfer efficiency.

87.3 Equipment

The equipment necessary to carry out electrostatic spraying consists of:

A. A power supply.
B. A manual electrostatic spraygun.

The DeVilbiss Company Ltd manufacture a power supply unit known as the BFA–800. This unit has the voltage adjustable in two ranges: 0.60 kV manual application or 0.90 kV automatic application. LED displays provide continuous monitoring of kilovolt and microamp readings. The unit has modular circuitry

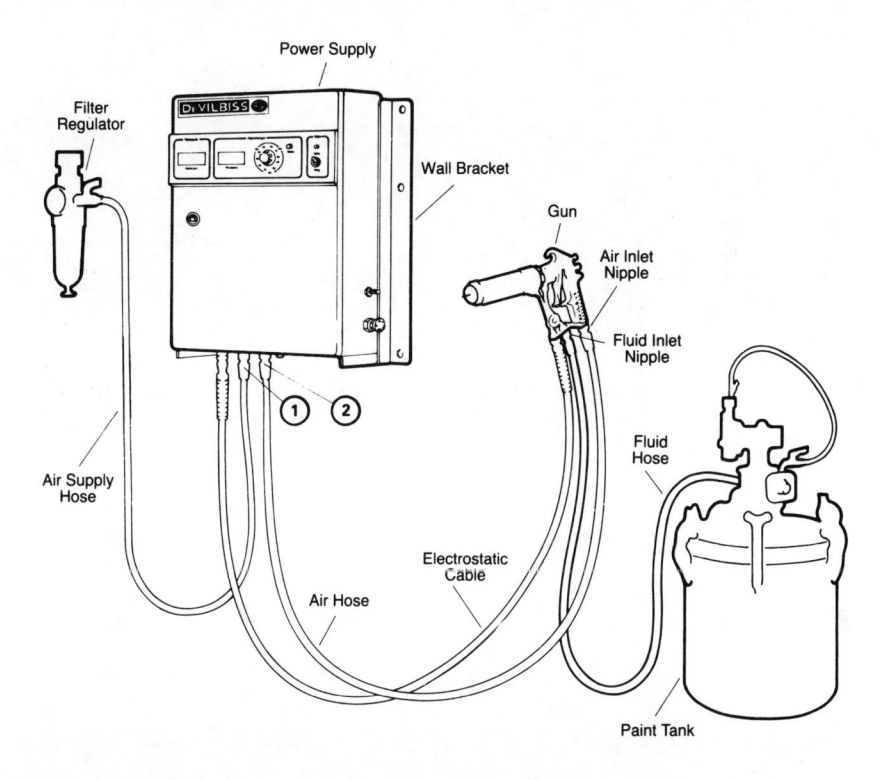

Fig. 42. DeVilbiss manual electrostatic liquid system layout

87.3 Vehicle Painter's Notes

for easy servicing, and built-in protection against live voltage variations. To operate with this power supply DeVilbiss Company Ltd market a spraygun known as the BFL–800 manual electrostatic spraygun. This unit is reliable and safe to use. Designed for ease of cleaning and maintenance, it has a high fluid flow rate and both round and fan spray caps can be used. The charged probe may be touched to any earthed objects without danger of sparking. For operator safety current will only flow to the gun when the trigger is operated.

The technology advance that this type of equipment offers will certainly ensure that it is more widely used in the finishing and refinishing industry.

Figure 42 shows a complete manual liquid system.

88
Metamerism

> Definition of metamerism – how it affects colour matching.

It is not always remembered that colour seen by the eye changes with the nature of the illumination. That is to say, colour as seen by normal vision depends not only on the spectral reflectance of the sample but on the quality of the light illuminating the sample. It is a common experience that the appearance of a colour is completely distorted in monochromatic sodium light, for instance, or in the light of a mercury vapour lamp. Even in the normal range of daylight, or between daylight and artificial (tungsten) light, pigments that match under one source of illumination may not match under another. This divergence is known as *metamerism*.

The likelihood of metamerism between different batches of paint is reduced if the same pigments are used in each batch. The occasion may arise, however, when the manufacturer is obliged to use pigments other than those in the normal formulation. In this case, careful formulation is necessary to keep metamerism within acceptable limits.

The maintenance of permanent colour standards against which each batch is compared is essential in paint manufacture in order to avoid long-term drift.

It is important for refinishers to understand metamerism as it can affect colour matching in the workshop, and a customer may view a vehicle in the light of a showroom which can alter the colour and highlight a repair.

89
Fault – Bleeding

> **89.1 Causes:**
> The main cause of this fault is contamination from underlying material, usually a red.
>
> **89.2 Prevention:**
> Test finish before applying colour and always ensure that spraying is carried out in the booth so that colour contamination does not occur.
>
> **89.3 Rectification:**
> In mild cases it can be rectified by using an inhibitor or sealer. More severe cases require stripping to bare metal.

This fault normally occurs with red, maroon or yellow finishes. It happens when the pigment of the original finish dissolves in the solvents of the refinish material and causes a discoloration.

89.1 Causes

A. Overspray from a colour prone to bleeding.
B. Improperly cleaned equipment.
C. Failure to seal off the original finish adequately.
D. Contamination of the undercoat system with material prone to bleeding.

89.2 Prevention

A. Do not allow spray dust from colour which may bleed to fall on other jobs.
B. Always clean all equipment thoroughly.
C. Test original finish before proceeding by applying a full wet coat of the colour to be used to a small flatted area. If the fault occurs, then seal back with a bleeding inhibitor as recommended by the paint manufacturer.
D. Never mix other products into the undercoat.

89.3 Rectification

In mild cases use a bleeding inhibitor and follow the manufacturers' recommendations carefully.

In a more severe case stripping to bare metal is the only sure way of eradicating this fault. Do not hesitate if in any doubt, as, if the condition is left it can show at a later date.

90
Fault – Paint Blister

90.1 Paint blister:
On a new vehicle this is usually a preparation fault, when phosphate salts are left in the grain boundary of the steel panelwork.

90.2 Spot blister rectification:
This can be easily achieved by grinding out the blemish and repainting as a spot repair.

90.3 Causes:
A list of causes of the spot blister fault.

90.4 General rectification:
By wet flatting panels and carefully observing the fault, decisions can be made as to whether to part refinish or strip the panel to bare metal.

90.5 Rusting:
When blisters occur on wing tops etc., it is often caused by the presence of water that has passed from the underside of the panel, and total rusting has occurred.

90.6 Prevention:
By use of underseal to protect the underside.

90.7 Rectification:
The only certain way is to cut away the rusted steel panel and weld in new material.

90.8 Blister types:
There are two categories of blister, contamination and insufficient paint film thickness.

90.9 Acceleration of blistering process:
Contaminates and film build will contribute.

90.10 Prevention:
A list of the necessary measures.

90.1 Paint blister

There are a number of faults that come under the heading of paint blisters. It can range from large blemishes arising up out of the film to the fine micro-blistering that is hardly detectable in the paint film.

The large blister that forms up in the paint film with a rust point on the surface is due to moisture penetrating the film and a process known as osmosis taking place. This fault is due to phosphate contamination at the point of manufacture. At all the major car manufacturers the steel body of the vehicle is degreased and treated with a phosphate solution to clean the grain boundaries of the steel and prepare it for primer coating. This phosphate solution is washed off with de-mineralised water and then dried in a hot air booth set up on the line. Sometimes, however, not all of the phosphate is cleaned out of the grain of the steel and as a result it remains below the surface of the subsequent primer and colour coats. As all paint is a semi-permeable membrane it allows moisture to penetrate through it. As this happens the moisture eventually reaches the pocket of phosphate and becomes a salt solution. Then the process of osmosis begins.

The salt solution is the stronger of the two (that is, the outside moisture and itself) and as they are separated by a semi-permeable membrane the stronger solution draws in more moisture, and the flow continues from the weaker to the stronger. As a result the paint film slowly erupts upward until the surface is finally disturbed. Then all surface protection has broken down and the moisture reaches down to the steel panel and the rusting out procedure commences.

Figure 43 shows the stages that occur.

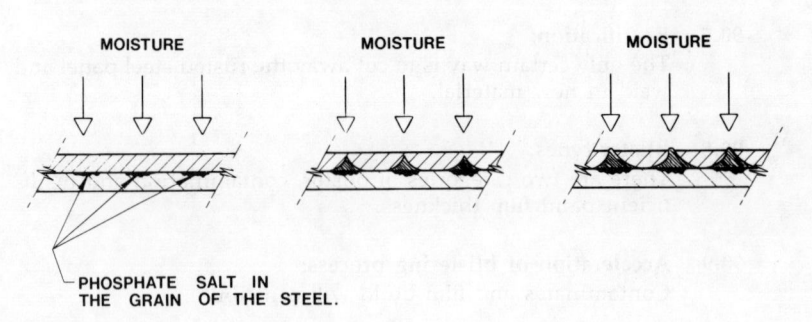

Fig. 43. Osmosis

90.2 Spot blister rectification

Using a DA sander fitted with a 240 grit disc clean out the rust spot and feather out accordingly. Watch for tell-tale trails of rust that by capillary action can be drawn along beneath the paint to another rust blemish. If this is the case then strip the whole panel, either mechanically or using a chemical stripper. If, however, it is isolated, then after feathering back build up with primer and fillers and then block back with 600 wet or dry and finish out in the normal way.

Even the best paint films are permeable to water vapour. When vapour penetrates the film it may set up forces sufficient to weaken the inter-coat adhesion between the various coats, or the adhesion of a whole paint structure to the underlying metal. As a result blisters containing water may form.

90.3 Causes

- A. Surface contaminated before painting, as already described.
- B. Insufficient undercoating.
- C. Insufficient top coat.
- D. Non-approved thinner.
- E. Non-approved paint system.
- F. Exposure of the finish to wet weather or conditions of high humidity before it is really hard.
- G. Continuous exposure to severe wet weather and conditions of high humidity.

90.4 General rectification

- A. Flat thoroughly until all signs of the blistering have been removed, e.g. if blistering is confined to the top coat, only that layer needs to be completely removed. There is always the danger of flatting residues obliterating the base of the blister and the moisture held in the blister may have caused deterioration of the paintwork beneath it. The only cure is:
- B. Strip to bare metal either mechanically or using a chemical stripper. Use a phosphoric acid metal cleaner and metal conditioner. Assist the action with wire wool. Then repaint as usual.

90.5 Rusting

The presence of a solitary blister or a group of blisters of varying sizes on wing tops or over headlight surrounds, etc., is usually an indication of metal deterioration by way of corrosion from the underside. Minute holes in the metal allow water to enter the paint system and blistering develops.

90.6 Prevention

Protect the underside of the vehicle by thorough undersealing.

90.7 Rectification

A. Filling, stopping and then repainting will give only a temporary solution to the problem, and in fact the problem will recur very quickly.

B. The only certain way to stop this is to replace the steel with a new piece welded into position and then correctly undersealed, and the top surface repainted as normal.

This highlights the need to constantly wash off under a vehicle with a hose to remove the road dirt and salt that gets caked up under the vehicle in normal use.

90.8 Blister types

Blistering may be divided into two categories:

A. Blistering resulting from contamination during the refinishing process or from failure to remove contaminants already present. The source of contamination may sometimes be easily identifiable for the blistering may follow the pattern of the contaminate, e.g., finger prints or wiping marks.

B. Blistering through other causes e.g., insufficient undercoat and or top coat, use of incorrect thinner. The blistering may take the form widely known as micro-blistering, although the true definition of micro-blistering is that it requires a low-powered magnification to be seen with the naked eye. It may manifest itself as loss of gloss or humidity blistering. All paint systems will absorb water during periods of rainfall and high humidity, drying out again during dry weather. With prolonged exposure to severe wet weather and conditions of high humidity a great deal of water may be absorbed. Little or no evaporation can take place and the paint has little chance to dry out. Eventually the point may be reached when sufficient water has penetrated the film to create enough pressure to lift the paint from the substrate, either from the undercoat or from original finish or bare metal, and form water-filled blisters.

90.9 Acceleration of blistering process

The process will be accelerated if:

A. Contaminates are present in the paint film, e.g. soluble salts from the flatting water, or perspiration from hands.

B. There is insufficient thickness of paint, either undercoat or topcoat. The first sign of failure may be micro-blistering on the feather edge of primer surfacers, where the thickness of the primer surfacer is very low. It is good practice to seal the edge with an extra coat of colour before proceeding with normal colour application.

Every extra coat of paint increases resistance to blistering, but high paint film thickness can bring other problems, like prolonged hardening time and crazing or checking. Paint manufacturers formulate their materials to give a high degree of blister resistance at economical film thickness, and the refinisher should make every effort to ensure that the manufacturers' recommendations of film thickness are strictly observed.

C. The protective upper surface of the paint film is damaged in any way, thus permitting earlier ingress of moisture into the paint system.
D. The vehicle is poorly maintained. Immediately a vehicle goes into service it is subject to attack from grit and dirt. These damage the top surface of the paint system, making moisture penetration easier. Regular washing and cleaning will remove the dirt and road grit and certain cleaning agents impart a protective film to the surface, but in time surface damage will be such that moisture will have an easier way in. Not all cleaning materials are suitable, some can actually accelerate the deterioration of the finish.
E. The vehicle is stored in the open or in a poorly ventilated garage for long periods. The best way to dry out paintwork is to run the vehicle on a dry day.

90.10 Prevention

To produce paintwork of maximum blister resistance, observation of the following points is of the greatest importance:

A. All surfaces are clean and free from contamination.
B. Flatting water is changed frequently and all flatting residues are removed.
C. Bare hands are kept off the workpiece.
D. The paintshop is maintained at the correct operating temperature, which is 68 to 72° F.
E. The compressed air is clean and free from any contamination.
F. The vehicle is allowed to reach the shop temperature before any application of paint.
G. The recommended thickness or film weight of the undercoat and topcoat are applied.
H. Adequate flash off and drying times are given.
I. The paintwork is allowed to dry out thoroughly before exposure to wet and very humid conditions.

To add to the points raised, an additional item of equipment is most useful to help in the eradication of micro-blistering. That is the use of infra-red drying lamps, and this subject is covered in Chapter 38. However, it is worth noting here that 3 micron medium wave infra-red does have a marked effect on the eradication of moisture from a semi-permeable membrane, and can give great assistance during the painting process.

91
Fault – Blooming or Blushing

> **91.1 Blooming or blushing:**
> This fault is usually associated with damp or cold weather. It manifests itself as a milky finish in the colour coat.
>
> **91.2 Causes:**
> Some of the main causes are excessive humidity, use of a poor quality thinner or use of a thinner which is too fast.
>
> **91.3 Prevention:**
> To prevent the fault it is necessary to heat the spraybooth and use the correct thinners.
>
> **91.4 Rectification:**
> To rectify the fault either compound and polish when the paint film is through dry or warm up the booth and spray a further single coat of colour.

91.1 Blooming or blushing

This fault is usually associated with the colder weather and normally occurs in refinishing shops that are poorly equipped or have insufficient heating. The fault appears as a milky film over the surface of the paint at the application time and during flash off. Condensation in the atmosphere stops the solvents evaporating from the surface of the film into the air, and subsequently they precipitate or fall back onto the paint surface.

91.2 Causes

The causes for this fault are:

A. Excessive humidity levels.
B. Use of poor quality thinner.
C. Use of a thinner which is too fast.
D. A cold and draughty paintshop.
E. Poor air movement and/or lack of heat in the spraybooth.

91.3 Prevention

To prevent the fault:

A. Ensure the paintshop or booth are heated to the correct temperature for spraying (20–21° C or 68–72° F).
B. Use the correct thinner recommended by the manufacturer.
C. Use a slower thinner.

91.4 Rectification

If the fault is slight then after through drying the surface can be compounded and polished and this will remove the bloom.

If the fault is more severe, then it can be rectified by spraying another colour coat, after the booth has been heated. It is not good practice to spray neat thinners over the film as this will load up the solvent into the surface.

In very severe cases when water droplets have been entrapped, allow a through dry and wet flat the area. Then respray in the normal way.

When blushing is seen on the colour coats it may well have appeared on the undercoat system, but gone unseen. If this is the case then blistering will occur and inter-coat adhesion will occur at a later date. The point here is to ensure that conditions in the paintshop or booth are correct.

92
Fault – Bronzing

> **92.1 Bronzing:**
> This defect is peculiar to certain colours and is caused by a pigment defect.
>
> **92.2 Causes:**
> Mixing formula not correctly followed or hot spraying of certain colours.
>
> **92.3 Prevention:**
> Action required to prevent the fault.
>
> **92.4 Rectification:**
> Procedure to quickly rectify the fault.

92.1 Bronzing

This defect is peculiar to certain blue, maroon and black pigments. It is formed by a loosely adhering pigment layer in the surface which is slightly different in colour from the original paint, and which imparts a metallic sheen to the surface.

92.2 Causes

A. Certain pigments will show slight bronzing in any paint vehicle, and the painter has no control over this condition.
B. Recommended mixing formula not followed. Certain pigments will show bad bronzing if an upper limit is exceeded in the paint.
C. Hot spraying of some reds or maroons.

92.3 Prevention

A. Follow the recommended formula for colour mixing.
B. Some reds and maroons may have to be sprayed cold.

92.4 Rectification

A. Light hand polishing with a mild liquid polish will remove the bronze. Frequent weathering and an occasional polish will maintain a good appearance.
B. In severe cases, wet flat and respray the colour.

93
Fault – Cobwebbing

93.1 Cobwebbing:
Failure to atomise correctly – colour streams.

93.2 Causes:
Use of old paint that has thickened. Incorrect pressure (too high). Incorrect viscosity (too thick). Wrong thinners. Failure to allow paint to come to shop temperature.

93.3 Rectification:
Flat and respray – ensure gun test before working.

93.1 Cobwebbing

This fault occurs where the paint fails to atomise correctly and the colour streams into a long filament that is forced onto the surface to be painted. It looks like a fine cobweb, hence the name.

93.2 Causes

The causes of cobwebbing are:

A. The use of old paint that has 'thickened'.
B. An incorrect spraying pressure which is too high.
C. An incorrect viscosity where the material is too thick.
D. Use of incorrect thinners.
E. Failure to allow the paint to come up to workshop temperature.

93.3 Rectification

As soon as the fault is spotted stop the application immediately. It is necessary to allow the cobwebbing to through dry and then wet flat the material off the panel. This is a demonstration of how important it is to test the gun on a test panel in the spraybooth.

Working temperatures are very important, not only in the booth but in the shop generally. Both the temperature of the paint and the surface temperature of the panel to be sprayed contribute to the success of the overall job. The temperature window is between 68° F and 72° F (20–21° C).

94
Fault – Crazing

> **94.1 Crazing:**
> The description of the fault known as crazing.
>
> **94.2 Causes:**
> The main causes of the fault and the reason for the occurrence in the paint film.
>
> **94.3 Rectification:**
> The correct method of dealing with both surface and full depth crazing.

94.1 Crazing

The fault known as crazing gives a crazy paving pattern in the paint film. There are two types:

- **A.** Surface or light crazing.
- **B.** Full depth crazing.

94.2 Causes

The causes of this fault are generally as follows:

- **A.** Ageing of a paint film.
- **B.** The sandwiching of cellulose and synthetic finishes or undercoats.
- **C.** Using a cellulose stopper under an SR synthetic surfacer.
- **D.** Excessive delay in applying the second or third coat of a synthetic finish. In other words, applying colour during the critical recoat period.
- **E.** Applying colour to a paint film with a total build of in excess of 12 thou or 300 microns.
- **F.** Failure to bring an acrylic painted vehicle up to shop temperature before applying paint.
- **G.** Heavy wet coats will aggravate this fault.

94.3 Rectification

- **A.** Surface crazing. The use of P 1200 wet and dry on the panel, followed by compound and polish, should cure this problem. It is unlikely to recur.
- **B.** Deep crazing. When this occurs the only way forward is to strip the panel back to bare metal and commence refinishing.

95
Fault – Dry Spray

> **95.1 Dry spray:**
> Paint arriving on surface in dry powder condition.
>
> **95.2 Causes:**
> Spray gun too far from panel. Spray gun pass too fast. Cheap thinner. Too fast thinner. Air pressure too high. Wrong viscosity.
>
> **95.3 Rectification:**
> In primer – wash off. In colour – wet flat 1000 – 1200 – refinish – polish – depends on condition.

95.1 Dry spray

The definition of dry spray is where paint arrives on the panel in a dry powdery condition.

95.2 Causes

The causes of this fault are as follows:

A. The spraygun is too far from the panel.
B. The spraygun pass speed is too fast.
C. A 'cheap' or incorrectly balanced thinner has been used.
D. A thinner that is too fast for the job has been used.
E. The air pressure at the gun is too high.
F. The viscosity of the paint is incorrect.
G. The spraybooth conditions are too hot.

95.3 Rectification

If this fault occurs in the primer coats then it can be washed off with solvent immediately. It is preferable to do this as, if it is left on the job, and oversprayed with more primer or filler, then porosity can occur which will cause further problems later, especially if the vehicle is placed in a low bake oven. In colour, however, it is not possible to wash off, and therefore the material should be allowed to dry out and then the surface should be well flatted back. The panel can then be refinished out correctly.

95.3 Vehicle Painter's Notes

Sometimes if the dry spray is fairly minor, then the panel can be flatted back with P 1200 paper wet and dry and then polished with compound and a liquid cutting polish such as T Cut.

The skill is in deciding what course of action to take with each particular job. Although flatting and polishing may be the answer it is just as quick to wet flat back and coat up with colour.

96
Fault – Floating and Flooding in Metallic

> **96.1 Floating and flooding:**
> Pigment or metal flakes float in solvent on the surface of the job giving a dark and patchy appearance.
>
> **96.2 Causes:**
> Too much material applied giving a heavy wet film. Material overthinned, or incorrect spray technique.
>
> **96.3 Rectification:**
> Flat and respray.

96.1 Floating and flooding

These faults occur when too much material has been applied to a panel and large patches of dark wet appearance occur. The pigment or metallic particles actually float in the solvent and behave in a random manner.

96.2 Causes

Too much material has been applied, giving a heavy wet film. The material may have been overthinned or the spraying air pressure was incorrect (too low). The spray technique could also have been the cause.

96.3 Rectification

If the floating and flooding is not too severe then it can be rectified by dry spraying more colour over the fault and allowing the solvent to soak up into the dry spray, rather like ink into blotting paper. After a thorough dry through the colour can be either compounded and polished or wet flatted back with P 600 wet and dry and resprayed.

If the situation is more serious it is better to allow the solvent to evaporate away, and then allow a full through dry to take place and finally wet flat the panel back with P 600 wet and dry. The panel may then be resprayed as normal.

96.3 Vehicle Painter's Notes

The operator has to be careful that solvent entrapment does not occur in this condition, as only either flatting right back to substrate primer or stripping to bare metal will cure that particular problem.

Again, the advice and watchword is care at every stage so that this type of fault is never allowed to occur.

97
Fault – Metallic Air Pressure

> **97.1 Air pressure:**
> The pressure of air is a major factor in the spraying of metallics. Variations will alter the appearance and texture of the job.
>
> **97.2 Air pressure too high:**
> Too high a pressure will result in loss of solvent and a dry, silvery appearance.
>
> **97.3 Air pressure too low:**
> This will cause an orange peel effect, and give a darker appearance.

97.1 Air pressure

When refinishing metallic paint films the whole process has to be carefully monitored as there are many more constraints on the refinisher. Errors that may occur are greatly magnified because of the metallic inclusions in the paint film.

As an example, if the air pressure is too low in a straight colour, then other than an orange peel effect that can be rectified by wet flatting, and the film remaining slightly soft, the appearance can be enhanced well enough. However, in metallic finishes, the lower air pressure will mean a difference in the colour, i.e., it will be darker, as well as the orange peel and the soft paint film. Obviously a mismatch cannot be flatted back and polished because the colour is inconsistent with the original colour.

97.2 Air pressure too high

A. Will cause a loss of solvent in the atmosphere and a rapid dry of the paint which will appear 'dusty' on the painted surface.
B. Will cause the paint particles to 'bounce' off the surface to be painted and collide with the stream of paint coming from the gun. This will cause overspray 'fog'.
C. Will cause a high overspray content in the booth environment which will settle back on the surface, causing a 'sleepy' appearance.

97.3 Air pressure too low

A. Will cause an orange peel effect that cannot be rectified by wet flatting and polishing.
B. Will cause a darker appearance to the colour coat and therefore a mismatch against the original finish.
C. Will cause a heavy build of paint that will remain soft, and also could lead to solvent trapping that will cause pinholing.

Use the correct pressure and aim for a good gloss from the gun.

98
Fault – Peeling

98.1 Peeling:
This fault occurs where preparation has been poor, and manifests itself as the paint film simply peeling off the substrate.

98.2 Causes:
Some of the main causes are the use of incorrect primer or undercoat, poor application of the primer and excessive colour application.

98.3 Prevention:
To prevent this fault occurring, correct procedures must be followed throughout the painting process.

98.4 Rectification:
The only rectification is to strip and respray.

98.1 Peeling

This fault generally appears when preparation has been poor, or the operator has not followed the correct procedure. Paint flakes away and peels off from the substrate or even down to the bare metal panel.

98.2 Causes

The causes of this fault are:

A. Contamination of the surface by wax, silicone, soap, detergent or stearate powder (this is a lubricant used in dry flatting paper or DA discs).
B. The use of incorrect primer or undercoat.
C. Poor flatting at any given stage.
D. Poor or incorrect application of primer.
E. The primer left too long before recoating.
F. The use of cheap or unbalanced thinners.
G. Failure to use self etch primer, especially on aluminium panels or GRP.
H. Masking too soon over soft paint.
I. Removing masking tape after paint is through dry.
J. Excessive application of colour coats.

98.3 Prevention

It is absolutely necessary to follow correct procedures and use the correct materials and thinners for the job in hand. Any deviation can lead to this fault occurring.

98.4 Rectification

The only rectification for this fault is to strip to bare metal and commence the full paint procedure once again.

99
Fault – Pinholing

> **99.1 Causes:**
> There are five main reasons why this fault occurs, and the most common are application of colour coat over dry sprayed primer filler and areas of body filler not correctly sealed off.
>
> **99.2 Prevention:**
> A list of preventive measures to ensure that the fault does not occur.
>
> **99.3 Rectification:**
> The simple rectification procedure of flatting out and refinishing or in severe cases stripping to bare metal.

99.1 Causes

- A. Where there has been an application of colour coats over an undercoat or colour coat which has been dry sprayed.
- B. Air bubbles trapped during the application of lacquer when high viscosity, low air pressure and hot dry conditions combine to cause poor atomisation and rapid skin drying. The worst condition is when high air temperatures are combined with rapid air movement.
- C. Pinholes in existing finish not completely removed during the flatting operation before repainting.
- D. Poor knifing technique when using stopper.
- E. Areas of body filler or stopper not sealed off before painting.

99.2 Prevention

- A. Avoid dry spray in undercoat systems or first colour coats.
- B. Ensure correct spray viscosity and air pressure.
- C. Thoroughly flat off previous finish to remove any pinholing already present.
- D. Correct knifing of body filler or stoppers. Using the knife at an acute angle causes the material to roll under the knife forcing air bubbles in.
- E. Seal off areas of body filler or stopper with primer surfacer.

99.3 Vehicle Painter's Notes

99.3 Rectification

A. Minor pinholing confined to the colour coat may be removed by flatting with P 1200 wet and dry paper and then compounding.
B. In other cases wet or dry flat the affected paint to a depth ensuring complete elimination of the holes, and then refinish in the normal way. Or remove the affected paint and refinish from bare metal.

On no account attempt to bridge the pinholes with successive dry applications of primer surfacer. You cannot bury it.

C. Pinholes exposed after flatting body filler or stopper should be sealed off with a thin spread of stopper applied with the spreader held at 90° to the surface. This technique ensures that the stopper is forced well into the holes and that it is not dragged out again as the spreader moves on.

When pinholing is a persistent problem in a paintshop, check and adjust those conditions favouring rapid surface dry, i.e. paint viscosity, type of thinner, shop temperature and air movement.

With body fillers and catalysed stoppers improving both in quality and overall performance, this fault should be easily identified into the areas of paint application.

Care and discipline will ensure that this fault does not appear in the finished film.

100
Faults – Scratch Marks

> **100.1 Causes:**
> The main causes of the fault that is observed in the colour coat.
>
> **100.2 Rectification:**
> The rectification procedure of a scratched surface.

100.1 Causes

The causes of scratch marks showing up in a paint film are as follows:
- **A.** Use of too coarse a paper for the job.
- **B.** Insufficient drying time between coats.
- **C.** Too heavy an application of the undercoat system.
- **D.** Premature flatting of the colour coat.
- **E.** Colour repair on a solvent-sensitive film with too thin colour coat.

100.2 Rectification

Allow the film to dry through for at least two to three days. Then flat back the surface with P 800 or even P 1200 paper wet with soap. Then examine the film and compound and polish as normal. If this is not successful then flat back with P 800 wet and refinish the panel with two or three coats of colour. Avoid excessive film builds if at all possible.

101
Fault – Sinkage

> **101.1 Sinkage:**
> Drying proceeds and sinks into underlying imperfections.
>
> **101.2 Causes:**
> Excessive application of materials – no drying time – dry spray of undercoats allowing porosity – failure to spot prime – failure to stir heavy pigmented primer/filler – use of coarse papers – poor drying conditions – cold – no air flow.
>
> **101.3 Rectification:**
> Allow paint to harden 2/3 days – flat 1200 – compound – severe – flat and respray.

101.1 Sinkage

Sinkage occurs when drying of the total paint film proceeds and sinks into the substrate and underlying imperfections.

101.2 Causes

There are a number of causes of this fault and they are as follows:

A. Excessive application of materials.
B. Not sufficient time allowed between coats for through dry.
C. The dry spray of undercoats allowing porosity.
D. Failure to spot prime or primer build inadequate.
E. Failure to stir heavy pigmented primer fillers.
F. Use of too coarse a flatting paper.
G. Poor drying conditions.
H. Booth too cold.
I. Incorrect air flow in spraybooth.

101.3 Rectification

Allow the paint film to harden out for two to three days. Then flat the panel with P 1200 and compound the area.

If the sinkage is severe it is likely to return even after this course of action and therefore it is better to wet flat the surface back using a block and refinish the panel in colour.

Fault – Sinkage 101.3

If cellulose materials have been used take care not to load the surface up too much with solvent as the solvent penetration can cause further sinkage that will become visible in a longer period of time.

102
Fault – Solvent Popping

102.1 Solvent popping:
This fault occurs as solvent vapour breaks the surface of the film.

102.2 Causes:
The most usual causes listed.

102.3 Prevention:
The measures for prevention of the fault occurring.

102.4 Rectification:
To rectify either flat and refinish or strip.

102.1 Solvent popping

This condition is when bubbles spoil the surface of the finish. It is often incorrectly judged to be due to the formation of solvent vapour bubbles in the wet paint film, hence the term 'solvent pop'. Popping is usually the result of air bubbles, trapped during the application of the paint, being unable to escape because of a very quick film set up.

102.2 Causes

A. Hot dry conditions and/or excessive air movement.
B. Air pressure too low.
C. Insufficient flash off time between coats.
D. Application of excessively thick films.
E. The application of heat too soon after painting.
F. Heat source too hot or too close to the painted surface.
G. Use of cheap thinners or use of the incorrect thinners.

102.3 Prevention

A. Use the recommended thinners and add retarder during hot or very dry weather, but generally unnecessary if painting in a correctly functioning spraybooth.
B. Use the correct air pressure.

Fault – Solvent Popping 102.3–102.4

C. Do not apply thick coats and allow adequate flash off time between coats of material.
D. Allow ample flash off time before stoving.
E. Check oven temperature and adjust if necessary.
F. Check metal temperature of vehicle in oven.

102.4 Rectification

Either flat back the surface and refinish or strip to bare metal if the fault cannot be flatted out.

103
Fault – Striping in Metallic

103.1 Striping:
Shade variation with each pass of the gun.

103.2 Causes:
The causes of this fault are failing to overlap correctly, a defective spraygun, the gun too close to the panel, the spray fan too narrow, and failure to hold the gun at right angles to the job.

103.3 Rectification:
To alter the gun technique, then lightly dust coat the panel, or wet flat and finish out.

103.1 Striping in metallics

Striping occurs in metallic films only and is noticed as a shade variation at the pass of the gun over the workpiece.

103.2 Causes

A. Failure to overlap correctly with each pass. The overlap is determined by the character of each colour. For example, some very light blue metallics tend to 'float' more than dark and more dense maroons or greens. In the case of light blue it may be necessary to overlap the gun pass by up to 80%. The minimum overlap is 50% and therefore there is a 30% band in which to operate.

B. A defective spraygun or set up can also cause banding. Check the gun carefully for blocked air holes or imbalance in the spray fan.

C. The gun can be too close to the surface of the panel. Check carefully the distance (6 to 8 inches) and having set a distance, ensure that this is maintained consistently.

D. The spray fan may be too narrow. Check carefully that the spread is open and full.

E. The operator could be failing to hold the gun at right angles to the panel being coated. This will give a heavier application of colour either top or bottom, depending on the gun angle.

103.3 Rectification

This fault can be rectified by either dusting light coats of colour onto the surface to 'mop up' the areas of too much solvent, or the film can be left to thoroughly through dry, and then the surface can be wet flatted and then resprayed.

104
Fault – Trapped Air Popping

104.1 Causes:
Due to air entrapment in the paint film, the film skins over before the air can escape. It breaks through the surface and bubbles.

104.2 Prevention:
To prevent this the following should be observed: always use the correct thinner, the correct air pressure and apply thin light coats of material.

104.3 Rectification:
A light wet flat can cure the problem but often a strip to bare metal is necessary.

104.1 Causes

A fault that often causes confusion, as it is sometimes thought to be caused by solvent entrapment. Popping is usually the result of air bubbles being trapped during application by heavy film build which quickly sets up and traps in the air. Some of the major causes are as follows:

A. Air pressure too low.
B. Over-application or excessively thick films.
C. Hot dry conditions in the spraybooth.
D. Excessive air movement in the spraybooth.
E. Insufficient flash off between coats.
F. Application of heat too soon after painting.
G. Heat source too hot or too close to the panel.
H. Use of cheap or incorrect thinner.

104.2 Prevention

To prevent this condition the following should be observed:

A. Always use the manufacturers' recommended thinner.
B. Use the correct air pressure.
C. Apply thin light coats and allow adequate flash off between each coat.
D. Allow ample flash off time before commencing stoving schedule.

E. Check oven temperature and adjust if necessary.
F. Check metal temperature of vehicle in the oven.

104.3 Rectification

Examine the condition carefully, and if it is apparent that it has occurred in the colour coat only, a light wet flat with P 1200 followed by compounding and polishing will remove the fault.

In a worse case situation, wet flat the affected part to a depth where the popping has been totally eliminated, and then refinish in the normal way.

If the fault is in the full depth of the film, then strip to bare metal and refinish as normal.

On no account attempt to bridge the hole by spraying primer filler 'dry' to stop the imperfection.

Popping holes exposed after flatting should be sealed off with a fine spread of cellulose stopper, applied with the spreader held at 90° to the surface to avoid air entrapment and to ensure that the stopper does not get dragged out as the spreader moves on. If this fault is persistent in the shop then check conditions favouring rapid surface dry, i.e. paint viscosity, types of thinners, shop temperatures and air movements.